Cysteine - New Insights

Edited by Nina Filip

Published in London, United Kingdom

Cysteine - New Insights
http://dx.doi.org/10.5772/intechopen.111010
Edited by Nina Filip

Contributors
Aletta Millen, Alexandru Filip, Alin Constantin Pinzariu, Ann-Christin Niehoff, Bernhard Spengler, Bogdan Veliceasa, Catalina Filip, Cristiana Filip, Cristina Iancu, Daniela Matasariu, Demetra Gabriela Socolov, Dhaka Bhandari, Diana Zamosteanu, Gabriela Bordeianu, Gales Cristina, Gilles Frache, Ionela Lacramioara Serban, Luuk Van Oosten, Magdalena Cuciureanu, Marthe Verhaert, Mihaela Pertea, Minela Aida Maranduca, Nina Filip, Oana Viola Badulescu, Peter Verhaert, Raf Sciot, Remco Crefcoeur, Roxana Covali, Sooraj Baijnath

First published in London, United Kingdom, 2024 by IntechOpen
IntechOpen is the global imprint of INTECHOPEN LIMITED, registered in England and Wales, registration number: 11086078, 167-169 Great Portland Street, London, W1W 5PF, United Kingdom

British Library Cataloguing-in-Publication Data
A catalogue record for this book is available from the British Library

Additional hard and PDF copies can be obtained from orders@intechopen.com

Cysteine - New Insights
Edited by Nina Filip
p. cm.

This title is part of the Biochemistry Book Series, Volume 56
Topic: Chemical Biology
Series Editor: Andrei Surguchov
Topic Editors: Murat Şentürk and Deniz Ekinci

Print ISBN 978-0-85466-998-1
Online ISBN 978-0-85466-997-4
eBook (PDF) ISBN 978-0-85466-999-8
ISSN 2632-0983

For EU product safety concerns:
IN TECH d.o.o., Prolaz Marije Krucifikse Kozulić 3, 51000 Rijeka, Croatia,
info@intechopen.com or visit our website at intechopen.com.

IntechOpen Book Series
Biochemistry
Volume 56

Aims and Scope of the Series

Biochemistry, the study of chemical transformations occurring within living organisms, impacts all of the life sciences, from molecular crystallography and genetics, to ecology, medicine and population biology. Biochemistry studies macromolecules - proteins, nucleic acids, carbohydrates and lipids –their building blocks, structures, functions and interactions. Much of biochemistry is devoted to enzymes, proteins that catalyze chemical reactions, enzyme structures, mechanisms of action and their roles within cells. Biochemistry also studies small signaling molecules, coenzymes, inhibitors, vitamins and hormones, which play roles in the life process. Biochemical experimentation, besides coopting the methods of classical chemistry, e.g., chromatography, adopted new techniques, e.g., X-ray diffraction, electron microscopy, NMR, radioisotopes, and developed sophisticated microbial genetic tools, e.g., auxotroph mutants and their revertants, fermentation, etc. More recently, biochemistry embraced the 'big data' omics systems. Initial biochemical studies have been exclusively analytic: dissecting, purifying and examining individual components of a biological system; in exemplary words of Efraim Racker, (1913 –1991) "Don't waste clean thinking on dirty enzymes." Today, however, biochemistry is becoming more agglomerative and comprehensive, setting out to integrate and describe fully a particular biological system. The 'big data' metabolomics can define the complement of small molecules, e.g., in a soil or biofilm sample; proteomics can distinguish all the proteins comprising e.g., serum; metagenomics can identify all the genes in a complex environment e.g., the bovine rumen.

This Biochemistry Series will address both the current research on biomolecules, and the emerging trends with great promise.

Meet the Series Editor

Andrei Surguchov, Ph.D., joined Baylor College of Medicine, Houston, TX, as a faculty member in 1992, where he studied the mechanisms of the genetic control of lipid metabolism. At the University of Utah, his research interests focused on cloning new genes encoding retinal proteins. He studied molecular and cellular mechanisms of neurodegenerative diseases and retinal degeneration at Washington University, St. Louis. Currently, his research focuses on the structure-function relationship of proteins involved in neurodegeneration and ocular diseases. Andrei Surguchov is an Editor-in-Chief at Biochemistry Research International and Associate Editor in several biomedical journals.

Meet the Volume Editor

Nina Filip is an associate professor of biochemistry in the Faculty of Medicine, Grigore T. Popa University of Medicine and Pharmacy in Iasi. Representative areas of her research are the role of homocysteine in pathology, oxidative stress, and the implications of biogenic amines in human pathology. Rigorous academic training and passion for research have been the basis for the development of a successful academic career.

Contents

Preface

The study of amino acids is a vast and exciting field that can explore the molecular and cellular processes that underlie the functioning of living organisms. The amino acids cysteine and homocysteine contain the sulfur atom and are particularly important because they play a crucial role in various physiological processes.

Cysteine is involved in cellular redox homeostasis, being a key component of glutathione, an important antioxidant. Homocysteine is recognized as an important biomarker for several health conditions, and maintaining its levels within a normal range is considered beneficial for overall health.

The chapters in this volume will guide you step by step in discovering and understanding the fundamental concepts related to cysteine. Each chapter is dedicated to a specific aspect to provide a comprehensive summary of the latest insights into cysteine metabolism and its significance in different pathological conditions.

The structural similarity between homocysteine and cysteine makes the latter interesting to investigate for possible involvement in thrombophilia and deep vein thrombosis.

Throughout this book, you will discover a variety of information regarding cysteine and homocysteine metabolism, techniques, and methods of determination, as well as the implications of these amino acids in pathology. This information is essential for understanding the concepts and processes involving these amino acids.

In closing, we hope that this book will inspire and guide you on your journey to fascinating discoveries in the study of amino acids. Use your curiosity and passion to explore and understand the complex world of molecules and cellular processes that underpin life.

Nina Filip
Biochemistry,
Faculty of Medicine,
Grigore T. Popa University of Medicine and Pharmacy,
Iasi, Romania

Chapter 1

Introductory Chapter: General Aspects Regarding Cysteine and Homocysteine

Nina Filip

1. Introduction

Amino acids are fundamental building blocks of life, playing pivotal roles in various metabolic processes. These two amino acids, cysteine and homocysteine, contain the sulfur atom and are particularly important because they play a crucial role in various physiological processes [1, 2]. Cysteine was first isolated by Eugen Baumann, a German chemist, in 1884. Baumann was investigating the components of animal horn material when he identified cystine, the oxidized dimer form of cysteine [3]. Cysteine is involved in the synthesis of proteins, antioxidants, and important molecules such as glutathione, which help protect cells against oxidative stress. Modern techniques in protein engineering utilize cysteine residues to create more stable and functional proteins for therapeutic and industrial applications [4].

Homocysteine was first discovered by Vincent du Vigneaud, an American biochemist, during his research on the metabolic pathways of sulfur-containing amino acids. Du Vigneaud was awarded the Nobel Prize in Chemistry for his work on the synthesis of essential biological substances, including his research on methionine and homocysteine [5]. Kilmer McCully, an American pathologist, proposed the "homocysteine hypothesis" of arteriosclerosis. McCully's research showed that elevated levels of homocysteine could damage blood vessels, leading to cardiovascular diseases, such as heart attacks and strokes [6]. His work was initially controversial but later gained acceptance and led to a broader understanding of homocysteine's role in vascular health.

Homocysteine is a metabolite of methionine and is involved in methylation reactions and the synthesis of important molecules, such as DNA and neurotransmitters. However, elevated levels of homocysteine have been associated with several pathological processes, such as cardiovascular disease, neurodegenerative disorders, and impaired cognitive function. Additionally, research has shown that elevated homocysteine levels can contribute to the development of insulin resistance and diabetes. Furthermore, it has been suggested that the imbalance between cysteine and homocysteine levels can disrupt the redox balance in cells, leading to increased oxidative stress and cellular damage [7]. Advances in genetic research have identified mutations in genes involved in homocysteine metabolism (such as MTHFR) that can affect homocysteine levels and influence disease risk [8]. Personalized medicine approaches are being explored to address these genetic variations and optimize treatment strategies. These new insights highlight the intricate roles of cysteine and homocysteine in various pathological processes, shedding light on potential therapeutic targets for diseases associated with the dysregulation of these amino acids. New insights into the roles of cysteine and homocysteine in pathological processes have revealed their importance in various physiological functions, as well as their potential contribution to the development and progression of several diseases [9, 10].

2. Cysteine metabolism

Cysteine metabolism is a biochemical process that involves both synthesis and degradation of cysteine. Cysteine is an important amino acid for protein synthesis, detoxification, and various metabolic functions.

Cysteine synthesis occurs through the methionine transsulfuration pathway.

Methionine, an essential amino acid, is first converted to S-adenosylmethionine (SAM) by an ATP-dependent reaction.

SAM subsequently donates a methyl group in various methylation reactions, becoming S-adenosylhomocysteine (SAH), which is then hydrolyzed to homocysteine by S-adenosylhomocysteine hydrolase.

Homocysteine is converted to cystathionine by a reaction catalyzed by cystathionine β-synthase (CBS) with the help of serine. Cystathionine is then cleaved by cystathionine γ-lyase (CGL) to produce cysteine, α-ketobutyrate, and ammonia.

The activity of key enzymes like CBS and CGL is regulated by feedback mechanisms and can be influenced by the cellular levels of substrates and products.

Degradation of cysteine involves several pathways (**Figure 1**):

a. Cysteine dioxygenase (CDO) pathway: CDO converts cysteine to cysteine sulfinic acid, which can be further metabolized to taurine or undergo desulfurization to form pyruvate and sulfate. Taurine, an important molecule derived from cysteine sulfinic acid, plays significant roles in bile salt formation and cellular osmoregulation.

b. Desulfhydration pathway: cysteine can be directly converted to pyruvate by desulfurization, with the release of ammonia and hydrogen sulfide (H_2S) [11].

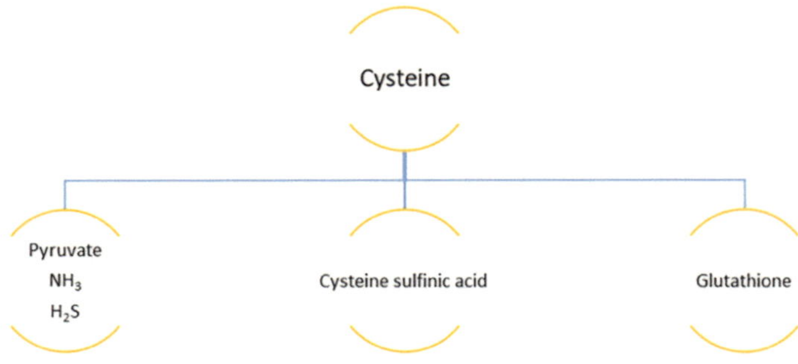

Figure 1.
Cysteine degradation.

c. Mercaptopyruvate pathway: cysteine aminotransferase catalyzes the transamination of cysteine with α-ketoglutarate to form 3-mercaptopyruvate which can be further converted by 3-mercaptopyruvate sulfurtransferase (MPST) to pyruvate and H_2S.

d. Glutathione pathway: cysteine is a precursor for glutathione (GSH), an important cellular antioxidant.

Cysteine degradation is a multifaceted process that integrates various metabolic pathways to maintain amino acid balance, detoxify sulfur compounds, produce important signaling molecules like hydrogen sulfide and contribute to the synthesis of biomolecules, such as taurine and glutathione.

3. Physiological roles and clinical significance

Cysteine plays several crucial roles in the body (**Figure 2**) [12]:

- Cysteine is a key component of many proteins, contributing to their structural stability through disulfide bonds [13].
- It is a precursor of glutathione, a major intracellular antioxidant that protects cells from oxidative stress [14].
- Cysteine participates in the detoxification of harmful substances through conjugation reactions in the liver [15].

Clinical relevance of cysteine [16].

Homocystinuria: a disorder caused by deficiencies in cystathionine-β-synthase, leading to the accumulation of homocysteine and associated complications, such as cardiovascular disease.

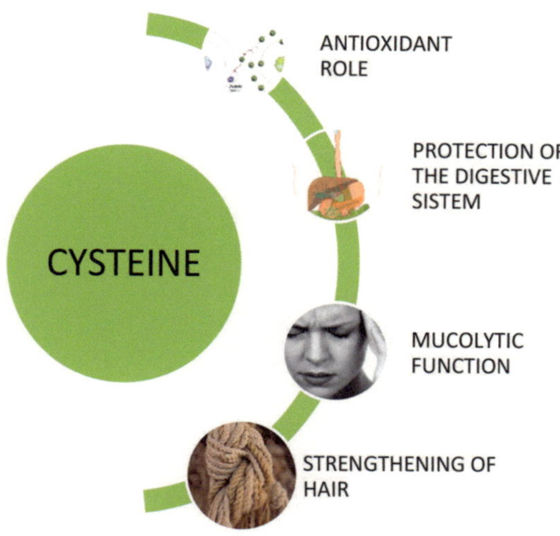

Figure 2.
The importance of cysteine in human health.

Cystinosis: a condition characterized by the accumulation of cystine (the oxidized dimer form of cysteine) in lysosomes, leading to kidney damage and other systemic problems.

Glutathione deficiency: can result from impaired cysteine metabolism, leading to increased oxidative stress and susceptibility to various diseases.

Homocysteine has significant clinical relevance:

- Cardiovascular health: elevated levels of homocysteine, known as hyperhomocysteinemia, are associated with an increased risk of cardiovascular diseases, including atherosclerosis and thrombosis [17].

- Neurological function: homocysteine levels are linked to cognitive function and neurological disorders, with high levels potentially contributing to neurodegenerative diseases [18].

- Gastrointestinal disorders: the association between hyperhomocysteinemia and inflammatory bowel disease, along with other autoimmune diseases, is supported by the role of homocysteine in promoting inflammation, oxidative stress, endothelial dysfunction, and immune dysregulation [19].

- Chronic renal diseases: elevated homocysteine (Hcy) levels have been observed in patients with chronic renal failure (CRF), those on dialysis, and even after a kidney transplant [19].

Today, homocysteine is recognized as an important biomarker for several health conditions, and maintaining its levels within a normal range is considered beneficial for overall health. The ongoing research aims to better understand the complex interactions between genetics, nutrition, and lifestyle factors in the regulation of homocysteine levels and their impact on health. Further studies are also focused on the potential therapeutic benefits of lowering homocysteine levels in various diseases.

Conflict of interest

The authors declare no conflict of interest.

Author details

Nina Filip
Faculty of Medicine, Department of Morpho-Functional Sciences (II), Discipline of Biochemistry, "Grigore T. Popa" University of Medicine and Pharmacy, Iasi, Romania

*Address all correspondence to: zamosteanu_nina@yahoo.com

References

[1] Muthuraman A, Ramesh M, Shaikh SA, Aswinprakash S, Jagadeesh D. Physiological and pathophysiological role of cysteine metabolism in human metabolic syndrome. Drug Metabolism Letters. 2021;**14**(3):177-192

[2] Papet I, Rémond D, Dardevet D, Mosoni L, Polakof S, Peyron MA, et al. Sulfur amino acids and skeletal muscle. In: Nutrition and Skeletal Muscle. Cambridge, Massachusetts, United States: Academic Press; 2019. pp. 335-363

[3] Berndt C, Buchanan BB, Lillig CH, Sies H. Thiol redox regulation: A brief historical overview. In: Redox Regulation of Differentiation and De-Differentiation. Boca Raton, Florida, United States: CRC Press; 2021. pp. 3-11

[4] Singh KR, Lee JK, Selvaraj C, Singh R, Li J, Kim SY, et al. Protein engineering approaches in the post-genomic era. Current Protein and Peptide Science. 2018;**19**(1):5-15

[5] Du Vigneaud V, Dyer HM. The chemistry and metabolism of the compounds of sulfur. Annual Review of Biochemistry. 1937;**6**(1):193-210

[6] Smith AD, Refsum H. Homocysteine– from disease biomarker to disease prevention. Journal of Internal Medicine. 2021;**290**(4):826-854

[7] Rehman T et al. Cysteine and homocysteine as biomarker of various diseases. Food Science & Nutrition. 2020;**8**(9):4696-4707

[8] Barati S, Fabrizio C, Strafella C, Cascella R, Caputo V, Megalizzi D, et al. Relationship between nutrition, lifestyle, and neurodegenerative disease: Lessons from ADH1B, CYP1A2 and MTHFR. Genes. 2022;**13**(8):1498

[9] Djuric D, Jakovljevic V, Zivkovic V, Srejovic I. Homocysteine and homocysteine-related compounds: An overview of the roles in the pathology of the cardiovascular and nervous systems. Canadian Journal of Physiology and Pharmacology. 2018;**96**(10):991-1003

[10] Kumar A, Palfrey HA, Pathak R, Kadowitz PJ, Gettys TW, Murthy SN. The metabolism and significance of homocysteine in nutrition and health. Nutrition & Metabolism. 2017;**14**:1-12

[11] Safrhansova L, Hlozkova K, Starkova J. Targeting amino acid metabolism in cancer. International Review of Cell and Molecular Biology. 2022;**373**:37-79

[12] Sameem B, Khan F, Niaz K. l-Cysteine. In: Nonvitamin and Nonmineral Nutritional Supplements. Cambridge, Massachusetts, United States: Academic Press; 2019. pp. 53-58

[13] Garrido Ruiz D, Sandoval-Perez A, Rangarajan AV, Gunderson EL, Jacobson MP. Cysteine oxidation in proteins: Structure, biophysics, and simulation. Biochemistry. 2022;**61**(20):2165-2176

[14] Raj Rai S, Bhattacharyya C, Sarkar A, Chakraborty S, Sircar E, Dutta S, et al. Glutathione: Role in oxidative/nitrosative stress, antioxidant defense, and treatments. ChemistrySelect. 2021;**6**(18):4566-4590

[15] Bonifácio VD, Pereira SA, Serpa J, Vicente JB. Cysteine metabolic circuitries: Druggable targets in

cancer. British Journal of Cancer. 2021;**124**(5):862-879

[16] Plaza C, Noelia MR, García-Galbis, and Rosa María Martínez-Espinosa. Effects of the usage of l-cysteine (l-Cys) on human health. Molecules. 2018;**23**(3):575

[17] Gospodarczyk A, Marczewski K, Gospodarczyk N, Widuch M, Tkocz M, Zalejska-Fiolka J. Homocysteine and cardiovascular disease—A current review. Wiadomosci Lekarskie. 2022;**75**:2862-2866

[18] Cordaro M, Siracusa R, Fusco R, Cuzzocrea S, Di Paola R, Impellizzeri D. Involvements of hyperhomocysteinemia in neurological disorders. Metabolites. 2021;**11**(1):37

[19] Al Mutairi F. Hyper-homocysteinemia: Clinical insights. Journal of Central Nervous System Disease. 2020;**12**:1179573520962230

Chapter 2

High Resolution Mass Spectrometry of Cystine-Containing Neuropeptides in Histological Sections of Human FFPE Tissue Banks

Peter Verhaert, Gilles Frache, Dhaka Bhandari, Luuk Van Oosten, Remco Crefcoeur, Bernhard Spengler, Marthe Verhaert, Aletta Millen, Sooraj Baijnath, Ann-Christin Niehoff and Raf Sciot

Abstract

Using our earlier developed protocol, mass spectrometry imaging of small endogenous peptides (and a selection of small metabolites) can be successfully performed directly in tissue sections of formaldehyde-fixed paraffin-embedded (FFPE) samples, such as those available in *Homo sapiens* biobanks. In analogy with immunohistochemistry (IHC) which employs antibodies as detection probes, this method was designated mass spectrometry histochemistry (MSHC) as it solely relies on (top-down) mass spectrometry for analyte detection. We demonstrate that MSHC enables the localization of cystine-containing neuropeptides in histological sections of human FFPE biobanked tissue and illustrate this on pituitary adenomas and non-diseased pituitary tissues archived for several years in an academic hospital pathology biobank. The instrumental setup consists of high-resolution mass spectrometers (several orbitrap systems and one dedicated hybrid TOF instrument) fitted with atmospheric pressure (AP) scanning matrix-assisted laser desorption/ionization MALDI. Currently, the best spatial resolution routinely achievable with such (MALDI) apparatus is 5 μm. The high mass spectrometric resolution obtained allows revealing the full isotope envelopes of the peptides. As such both reduced and oxidized cysteine-containing 'proteoforms' of e.g., the neurosecretory nonapeptides vasopressin and oxytocin can be visualized in biobanked FFPE tissue, demonstrating yet a novel application of MSHC.

Keywords: high-resolution mass spectrometry imaging, formaldehyde-fixed paraffin-embedded samples, *Homo sapiens*, biobanks, cystine-containing neuropeptides

1. Introduction

In a chapter of a book dedicated to the amino acid cysteine (Cys), it is probably redundant to emphasize the importance of this amino acid residue in protein/peptide biology. Yet, we would like to highlight one of the main reasons for our long-term interest in (endogenous) Cys-containing peptides. These peptides, often encoded by polypeptide precursor genes, are representatives of an extremely important class of secretory biomolecules, which fulfill crucial roles in intercellular communication. As such their functions in the organism belong to the (if not *the*) most essential and fundamental characteristics of 'life' [1, 2]. More than 40 years back, when the senior author of this chapter (P.V.) engaged in secretory peptide research, the most elegant way of demonstrating (and localizing) secretory biomolecules in tissues was by conducting immunohistochemistry-based experiments (see e.g. [3, 4]). Molecular characterization of newly discovered secretory peptides initially required laborious (if not heroic) extractions and chromatographic separations followed by tedious amino acid sequencing by automated Edman degradation (e.g. [5, 6]). With the advent of protein/peptide (tandem) mass spectrometry (MS), the identification and molecular characterization of secretory peptides became much more efficient (e.g. [7]), especially when it became evident that direct tissue matrix-assisted laser desorption/ionization (MALDI) MS enabled the immediate identification of secretory (neuro)peptides (e.g. [8, 9]). Several years later, the technology had developed thus far that scanning MALDI MS today allows *in situ*, i.e., on-tissue, analysis of secretory peptides, directly on histological sections of biological tissues, including samples that have been archived in tissue banks for a number of decades [10]. In this chapter, we illustrate the type of data that can be obtained from two different Cys-containing neuropeptides by modern high-resolution (HR) MS coupled to the latest atmospheric pressure scanning microprobe MALDI sources on well-documented samples from a hospital tissue bank.

We like to underline at this point that the sample histological sections we employed in this study all originated from formaldehyde-fixed paraffin-embedded tissue blocks that had been stored at ambient temperature in a hospital tissue bank for more than a few years. This remarkably contradicts several reports on MSI studies claiming that, without specific antigen retrieval steps (with their inherent risk of analyte delocalization and signal intensity reduction) and/or the use of special 'reactive' matrices for on-tissue chemical derivatization, top-down MS analysis of FFPE is destined to fail [11, 12]. Indeed, it has been suggested that formaldehyde fixation should preferably be avoided for MSI analyses [13].

2. Materials and methods

2.1 Samples

2.1.1 Sample origin and conservation

Homo sapiens pituitary-derived tissues were carefully selected from the large pathological collection at Leuven University Hospital FFPE biobank. The tissues were surgical specimens from patients who had undergone (anterior) pituitary tumor resection. All tissues had been fixed in formaldehyde and embedded in paraffin according to the hospital's standard operating procedure. Special attention was made to select tissue blocks that partly contained posterior pituitary nerve terminals, known to store the nonapeptides vasopressin and oxytocin, as validated by histochemical stains, including hematoxylin/eosin (H&E) and immunohistochemical (IHC) staining with nonapeptide antibodies (see below).

2.1.2 Sample processing

2.1.2.1 Sectioning

Paraffin blocks containing the tissue samples were trimmed and microtome sectioned at 5 μm thickness. For use in the orbitrap MS systems, the tissue sections were collected on regular [2.5 (or 2.6) × 7.5 cm] glass microscope slides and stored at room temperature until deparaffinization. For use in the TOF system, sections were mounted on conductive (indium tin oxide coated) slides.

2.1.2.2 Deparaffinization

Deparaffinization was performed essentially as described [1], the glass slides were sequentially immersed in a brief series of solvents, i.e., 2 times in xylene (100%) for 2 and 1 min respectively and two times in ethanol (absolute) for 1 min each. They were left to dry at room temperature for a minimum of 30 min before MALDI matrix application.

2.1.2.3 MALDI matrix coating

Before MALDI MS imaging, tissues were coated with MALDI matrix using automated pneumatic spraying or sublimation. For spray-coating, a solution of 2,5-dihydroxybenzoic acid (DHB, 50 mg/ml) was prepared in 50% acetonitrile, 50% H_2O containing 0.1% TFA. Two different spraying devices were employed yielding equal-quality matrix depositions. The sublimation instrument (see below) was operated using DHB powder (>99.0% pure) right out of the vial (Sigma-Aldrich #85707-1G-F).

2.1.2.3.1 M5 sprayer (HTX)

M5 sprayer settings were as follows: nozzle temperature: 75°C; nozzle height: 40 mm; flow rate: 0.1 ml/min; velocity: 1200 mm/min; track spacing: 1 mm; 3 passes; pressure: 10 psi; gas flow rate: 3 l/min; drying time: 2 sec.

2.1.2.3.2 Suncollect sprayer (Sunchrom)

Suncollect sprayer settings were as follows: z-distance: 25 mm; flow rate: 0.1 ml/min; velocity: 1200 mm/min; track spacing: 2 mm; 3 passes; pressure: 2.5 bar.

2.1.2.3.3 iMLayer matrix sublimation system (Shimadzu)

DHB was applied through an automated system as described [14] with sublimation resulting in a 2 μm matrix thickness (thinness) layer. To minimize matrix crystal size and assure analyte incorporation into the matrix crystals, a quick recrystallization step was performed (1.5 min) in a closed chamber at 75°C with a paper tissue with 500 μL MeOH:H_2O 5:1000.

2.2 Mass spectrometry

2.2.1 Ionization: MALDI

Different scanning atmospheric pressure (AP) MALDI sources were employed, depending on the mass spectrometer used.

2.2.1.1 AP-SMALDI

A Q-Exactive HF hybrid quadrupole/orbitrap mass spectrometer (ThermoFisher Scientific (Bremen) GmbH, Bremen, Germany) was connected to an AP-SMALDI[5] AF high-resolution MALDI imaging source (TransMIT GmbH, Giessen, Germany). The laser focus diameter on the sample was 5 µm and the sample was scanned with a 7 µm step size, resulting in an image pixel size of 5 µm × 7 µm.

2.2.1.2 AP/MALDI

Different generation orbitrap (ThermoFisher Scientific (Bremen, Germany and San Jose, CA USA) instruments were used in connection with an AP/MALDI (ng) UHR ion source (MassTech, Columbia, MD, USA). These comprised ion trap/orbitrap hybrid systems, including an LTQ Orbitrap Velos and an LTQ Orbitrap Velos Pro Elite; the single stage Exactive Plus, and the latest generation quadrupole/orbitrap Exploris 480. Laser energy was typical between around 5% (Nd-YAG laser), and laser focus was set below 10 µm for the Velos, Elite, and Exactive systems, to allow for a 10 × 10 μm^2 pixel size to be sampled in constant speed raster mode. The sample stage movement was programmed in the Target control software (MassTech) to be synchronized with the MS scan rate, to allow for, at least, 1 MS scan per pixel to be recorded.

2.2.1.3 Integrated AP MALDI

An iMScope TRIO (Shimadzu) ion-trap time-of-flight (IT-TOF) mass spectrometer Shimadzu Corporation (Kyoto, Japan) was also used, essentially as described before [14]. This instrument has a fully integrated atmospheric pressure MALDI source. Laser focus was <5 µm; laser intensity was set at 16.0 a.u., laser frequency was 1000 Hz and 200 shots were accumulated per pixel. For data acquisition, Imaging MS Solution 1.1 was used.

2.2.2 MS acquisition to imzML data format

2.2.2.1 QExactive HF

Mass spectra were acquired in positive-ion mode within a mass range of 500 to 2000 *m/z* and at mass resolution of 240,000 at 200 *m/z*. Internal lock-mass calibration on a DHB matrix cluster ion assured a mass accuracy of less than 2 ppm (root mean square error). Ion injection time was set to 500 ms, the S-lens level was maintained at 100 arbitrary units, and the ion transfer capillary temperature was 250°C. Raw MS data files (*.raw) and corresponding position files (*.udp) from the

AP-SMALDI source controller software were obtained as inputs for subsequent processing. The latest imzML converter ((c) University of Giessen and TransMIT GmbH) was employed for efficient data conversion. The converter utilizes an advanced algorithm transforming initial profile data into centroid data, reducing data dimensionality without loss of essential molecular information. Critical image dimension parameters including number of pixels (in x and y) and pixel size are automatically selected from the files, enhancing speed and accuracy of data processing, by minimizing manual intervention. Upon conversion process, the raw MS data are transformed into the imzML format. This generic format assures flexible data accessibility and facilitates downstream analysis, providing a foundation for robust interpretation and visualization of spatially-resolved molecular information [15].

2.2.2.2 Exactive Plus

The Exactive Plus orbitrap FT-MS system (ThermoFisher Scientific) was operated in positive-ion mode in a mass range of 150–1500 *m/z* at a mass resolution of 140,000. Ion transfer capillary temperature was set at 375°C.

2.2.2.3 Exploris 480

The Exploris 480 FT-MS system (ThermoFisher Scientific) was operated in positive-ion mode in a mass range of 400–1500 *m/z* at mass resolution of 60,000. Ion transfer capillary temperature was set at 375°C. Laser at 20% energy, 100 Hz. Max injection time 200 ms. AGC was on.

2.2.2.4 LTQ orbitrap Velos systems

We previously reported data obtained on early generation LTQ Orbitrap Velos PRO Elite and LTQ Orbitrap Velos systems, which were able to detect the secretory neuropeptides in question [16–19], but in this chapter we will concentrate on the younger ThermoFisher MS systems mentioned above.

2.2.2.5 iMScope TRIO

The iMScope TRIO IT-TOF mass spectrometer (Shimadzu) was operated in positive-ion mode in a mass range of 550–1500 *m/z*. Sample voltage was set at 3.5 kV; 200 laser shots were accumulated per pixel (laser frequency 1000 Hz).

To try and confirm the neuropeptide identity of the MSHC detected ion signals, tissue collision-induced dissociation (CID) MS/MS analyses were performed using *m/z* 1084.445 and 1007.443 ± 1.000 amu as precursor ion setting and adjusting the mass range to resp. 300–1100 *m/z* and 250–1100 *m/z*.

Spectra were processed using the Imaging Mass Solution software (v.1.1; Shimadzu).

2.3 Histochemistry

Two types of histochemical staining methods were used for light microscopic analysis of the tissue sections. These included classical H&E staining as well as immunohistochemistry (IHC).

2.3.1 Hematoxylin/eosin (H&E) staining

To facilitate histological recognition of the different cell and tissue types, an adjacent (serial) section of each paraffin block was stained with hematoxylin and eosin. For this the respective microscope slide containing a deparaffinized section was immersed in 70% ethanol (5 min), followed by a brief dip/wash in distilled water. Slides were then stained in Harris hematoxylin solution for 8 min, after which they were washed in running tap water for 5 min. Subsequently, sections were counter-stained in eosin-phloxine solution for 1 min, followed by conventional dehydration and cover-slipping; dehydration through 70% and 96% ethanol (2 changes of 5 min each), clearing steps in 2 changes of xylene (5 min each), and mounting with xylene based mounting medium.

2.3.2 Immunohistochemical (IHC) staining

Two different rabbit polyclonal antisera were used. Neuropeptide immunohistochemistry was performed using rabbit anti-vasopressin and anti-oxytocin polyclonal antibodies (a kind gift by Prof. L. Arckens; Laboratory for Neuroplasticity and Neuroproteomics, University of Leuven, Belgium). Both antisera had been used before in immunohistochemical studies to demonstrate vasopressin- and oxytocin-like immunoreactive peptides in neuronal tissues of various animal species [3]. Details of the immunohistochemical technique employed and antibody incubation parameters were described earlier [3, 4].

3. Results

To illustrate the specificity of our HRMS1 analyses, we can compare the experimentally obtained spectra with the calculated (theoretic) isotopic distribution of (a) vasopressin and (b) oxytocin respectively. **Figure 1** shows these both in oxidized (cystine, closed S—S ring) and reduced state (open disulfide ring; two free thiol S—H groups).

In the majority of the pixels sampled no neuropeptide ion signals are detected, which is in line with the known tissue distribution of the 2 nonapeptides focused on in this paper. Depending on the position where the tissue mass spectra are acquired, different (combinations of) ion signals are detected within the determined neuropeptide mass range. Some typical examples are shown in **Figure 2**. These comprise pixels (**Figure 2a**, **d** and **e**) that yield more or less complete neuropeptide ion isotope patterns; whereas in other pixels merely the most intense (monoisotopic) peaks are evident, such as for the (lower intensity) oxytocin ions (**Figure 2b** and **c**).

Using AP-SMALDI[5] AF—QE HF MSHC images were obtained at 7 μm lateral resolution. The distribution of vasopressin and oxytocin-related peptide ions is partly overlapping, but differential (**Figure 3** lower panel). Immunohistochemical staining demonstrated that vasopressin- as well as oxytocin-immunoreactive substances, are detected with a similar tissue distribution to the MSHC detected ions (**Figure 3** upper panel).

Using the AP/MALDI—Exploris 480 MSHC combination, images could be acquired at 5 μm lateral resolution on similar FFPE tissue from pituitary adenoma resection (**Figure 4**).

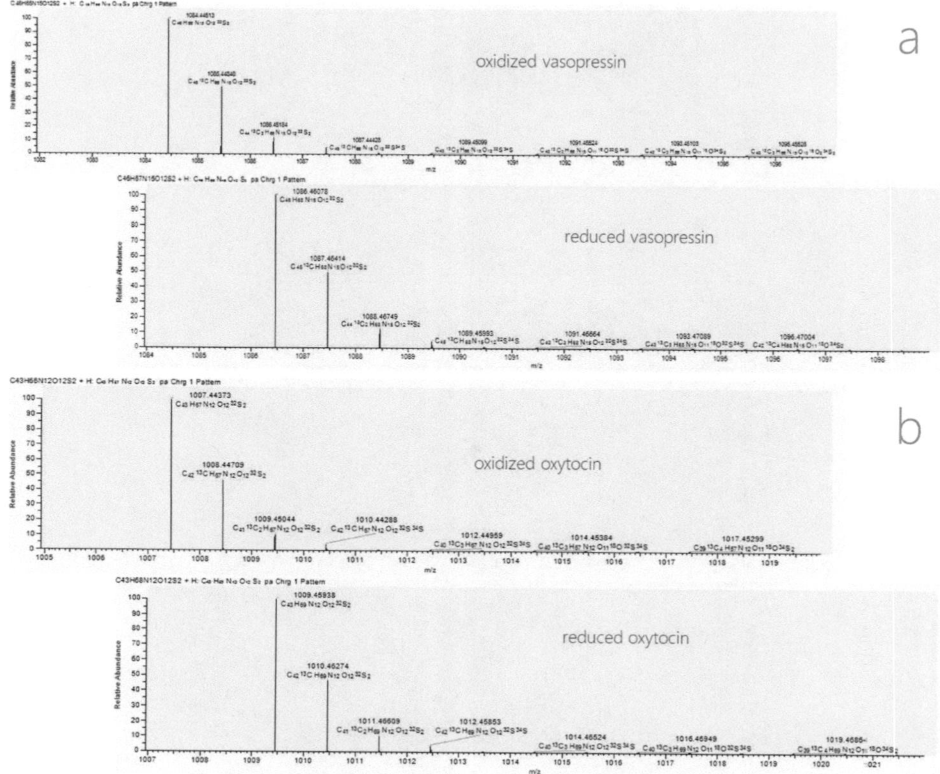

Figure 1.
Calculated (theoretical) isotopic distribution of vasopressin (a) and oxytocin (b). Note that, at observed m/z range, highest isotope peaks are invariably those with all C atoms as C12, followed by ions with one C13 atom, and, thirdly, ions with two C13 atoms in their composition.

Equally, ions representing both peptides are detected when the AP/MALDI source was mounted on earlier generation orbitrap systems, including the Exactive Plus (**Figure 2**, MSHC image not shown), and the 15 years-old LTQ Velos (previously published MSHC images [16, 17]).

Finally, also using the fully integrated iMScope TRIO can be used to generate images in the same spatial resolution range, as shown in **Figure 5**. The sample analyzed here was from a different pituitary adenoma patient than the one who donated the sample imaged in **Figures 3** and **4**. Immunohistochemistry of adjacent sections to the one imaged by MSHC yielded the expected images with no obvious distinction between the anti-vasopressin and anti-oxytocin staining (**Figure 5** upper panel).

CID tandem MS data were obtained both from the precursor ion at *m/z* 1084.445 and of that at *m/z* 1007.443 (**Figure 6**).

4. Discussion

As stated in our previous publication [18], many biologically active secretory peptides contain one or multiple disulfide bridges, in part due to/their contribution to the peptides' stability and activity. We concluded that the presence of Cys can

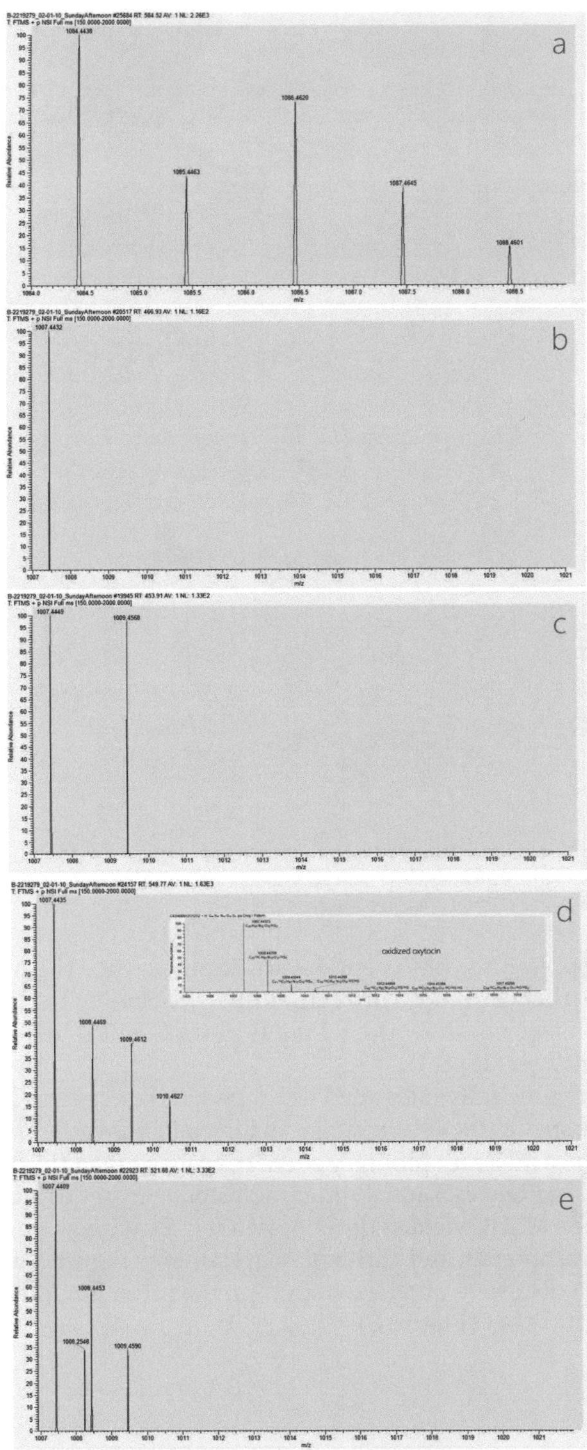

Figure 2.

Typical examples of on-tissue detected single-pixel isotopic patterns of vasopressin and oxytocin $[M + H]^+$ *ions (Exactive Plus data), showing different spectra depending on peptide quantities/signal intensities obtained. To facilitate comparison, insert in* ***Figure 2d*** *shows a theoretic isotope envelope of oxytocin with its cystine disulfide bridge formed.*

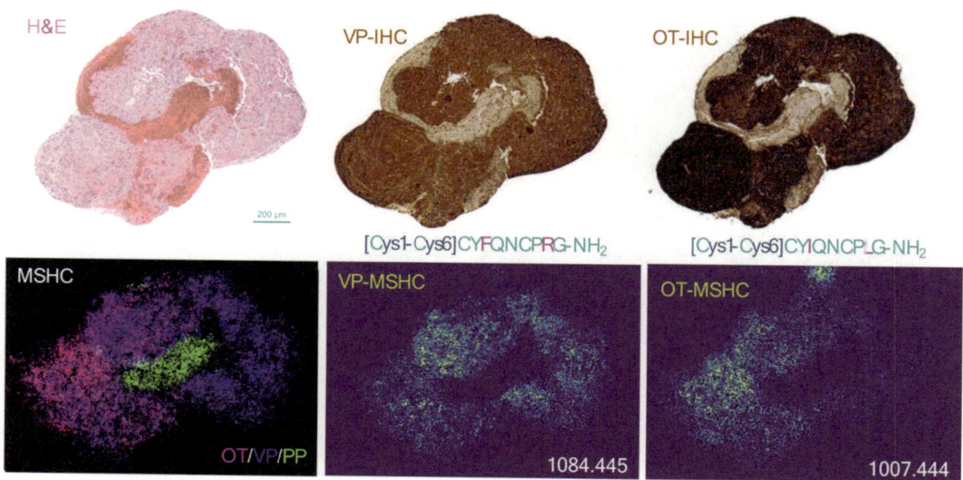

Figure 3.
AP-SMALDI[5] AF—QE HF MSHC images (7 × 7 µm² pixel size) generated in METASPACE (lower panel). Lower left image is an MSHC overlay of pixels containing protonated vasopressin (blue), protonated & sodiated oxytocin (resp. magenta & red), and protonated protoporphyrin (PP; $[C_{34}H_{34}N_4O_4 + H]^+$; precursor of heme; green). Upper panel shows serial sections of same paraffin block stained with H&E and IHC. Legend: H&E, hematoxylin/eosin; IHC, immunohistochemistry; MSHC, mass spectrometry histochemistry; OT, oxytocin; VP, vasopressin. Center lines show primary structures of VP (middle) and OT (right), with only 2/9 amino acid residues different.

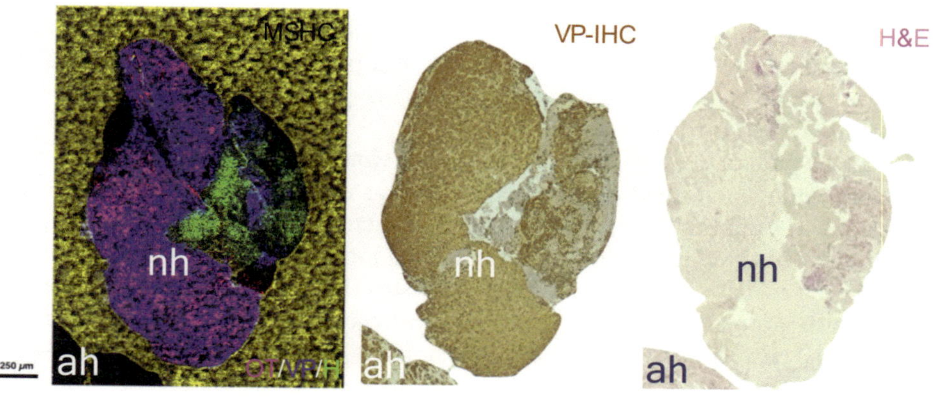

Figure 4.
METASPACE-generated MSHC image (left) obtained of biobanked human pituitary FFPE tissue with AP/MALDI 480 (5 × 5 µm² pixel size), with serial sections of the same paraffin block stained with anti-vasopressin antibodies (middle; VP-IHC) and hematoxylin/eosin (right; H&E). Abbreviations: ah, adenohypohysis; nh, neurohypophysis; H, heme b (m/z 616.16); OT, oxytocin (m/z 1007.44); VP, vasopressin (m/z 1084.44). Yellow pixels represent specific DHB matrix cluster ions, which appear to form predominantly outside of tissue.

be used as selection criterion for biologically important biomolecules in animal secretions. We calculated the prevalence of cysteines (6.91% of all residues) in the animal toxin sub-database of UniProt relative to the overall occurrence of cysteines in the whole UniProtKB/Swiss-Prot (release 2015_1) database (1.37% of all residues). We have recalculated these numbers based on the current versions of the respective databases (**Figure 7**). For this calculation, the latest version of the UniProt database (release 2024_1) was downloaded from https://www.uniprot.org/uniprotkb/statistics. The Tox-Prot database was derived from this using the

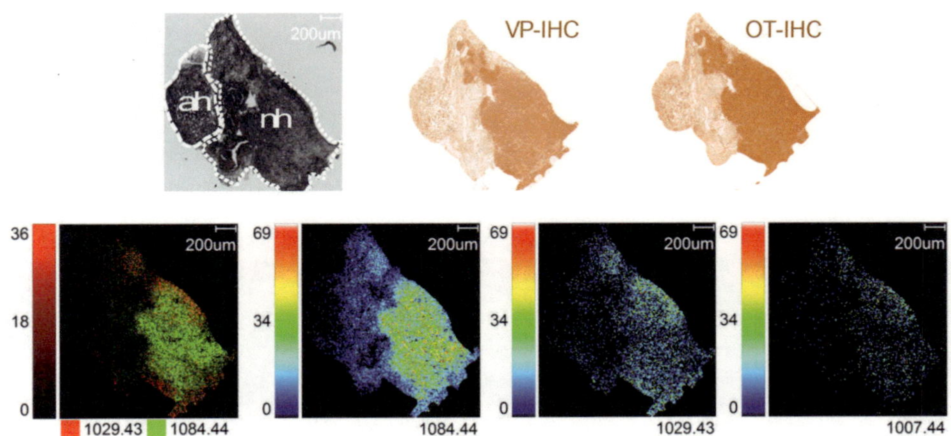

Figure 5.
Lower panel shows MSHC images (10 × 10 μm² pixel size) of resected human pituitary adenoma tissue archived in FFPE biobank, obtained after analysis with iMScope TRIO and generated using imaging mass solution software (v.1.1; Shimadzu). The numbers below represent selected m/z values of ions imaged, with far-left image being an overlay image of vasopressin ions in green and oxytocin ions in red. Numbers at the y-axis indicate ion intensity color scales. Upper panel left shows very same tissue section prior to MSHC analysis (and before DHB matrix application). Middle and right images are two serial adjacent sections immunostained with anti-vasopressin and anti-oxytocin antibodies. Legend: ah, adenohypophysis; nh, neurohypophysis; IHC, immunohistochemistry; OT, oxytocin; VP, vasopressin.

following query: "taxonomy_id:33208) AND ((cc_tissue_specificity:venom) OR (keyword:KW-0800)) AND (reviewed:true)" as described at https://www.uniprot.org/help/Toxins. We cannot but conclude that our previous observation [18] still stands, with 6.66% of all residues being Cys in Tox-prot, whereas only 1.39% of all residues are Cys in UniProt.

Since this publication, we have continued our work on yet another dimension to investigate endogenous peptides (Cys-containing as well as other secretory) in complex biological samples. We refer to the spatial dimension, and when applying mass spectrometry imaging technology directly to histological sections, one could refer to this approach as mass spectrometry histochemistry (MSHC). Whereas we originally studied fresh samples, minutes after prelevation/resection from the living organism [9, 20], in the past 7 years we have been focusing mainly on the type of material that is more easily (and ethically) accessible, i.e. formaldehyde-fixed, paraffin-embedded (FFPE) samples [10, 16, 17, 21].

We here focus on the data obtained for 2 known Cys-containing neurosecretory nonapeptides (vasopressin and oxytocin) which can easily be detected in *Homo sapiens* biobanked neuronal tissues [16, 17, 21].

The chemical structure of both vasopressin and oxytocin with their disulfide bridge formed is shown in **Figure 8**.

In this paper, we focus on MS1 data, which, on their own, obviously are debatably sufficient to validate the peptide identifications. However, we previously published that, using synthetic peptide analogs (both oxidized as well as reduced (open disulfide bridge) versions, tandem MS as well as ion mobility fully confirm, at least for vasopressin, our peptide identifications (see [21])).

Depending on the position of the tissue where mass spectra are acquired, different (combinations of) ion signals in the zoomed-in neuropeptide mass range are observed in the respective pixels, with typical examples shown in **Figure 2**. It shall be clear that

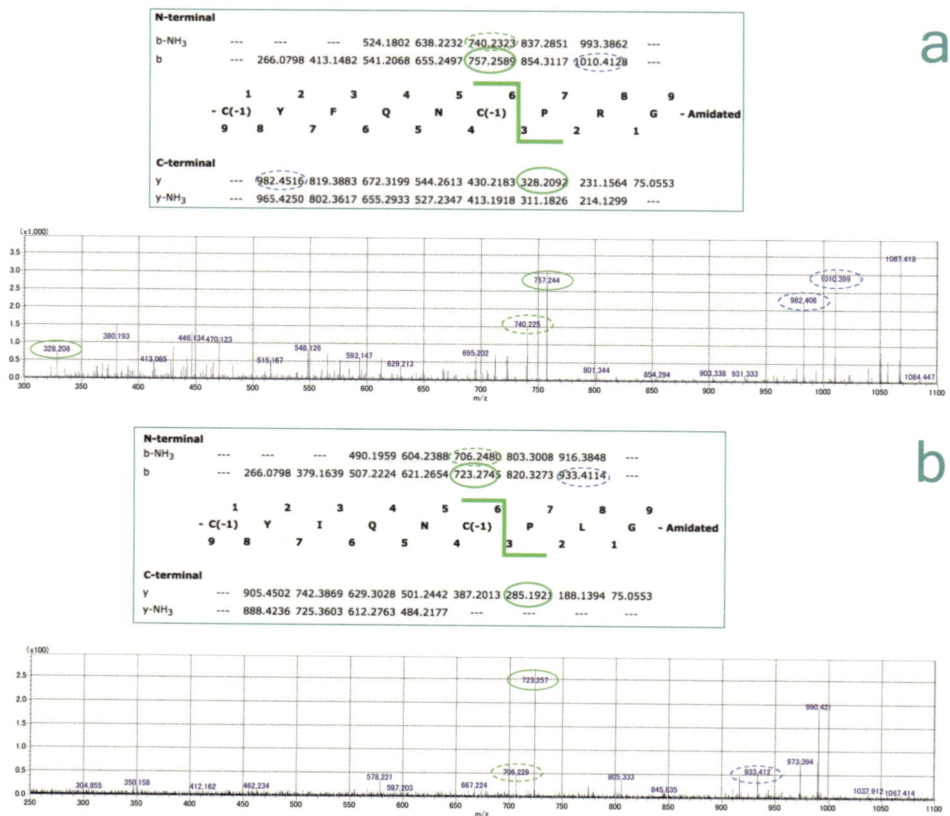

Figure 6.
On-tissue CID MS/MS spectra of precursor ions at m/z 1084.445 (a) and m/z 1007.443 (b). Respective inserts show vasopressin (a) and oxytocin (b) sequence with predicted fragment ion masses (online ProteinProspector MS Product output [https://prospector.ucsf.edu/prospector/cgi-bin/msform.cgi?form=msproduct]). Most fragile peptide bond (aminoterminally from Pro residue, carboxyterminally from disulfide bridge-linked cystine ring) is marked in green, yielding, for both nonapeptides, b6 and y3 as most dominant ions in the tandem MS spectrum.

not in all pixels, full isotope profiles are evident (**Figure 2b** and **c**). Indeed, it can be expected that in certain pixels some peptides may not be present at concentrations sufficient for the second (let alone third) isotopes to exceed the limit of detection. The fact, however, that the monoisotopic peaks detected exhibit the expected mass defect, and share the typical tissue distribution with the pixels that do contain enough neuropeptide ion signal for multiple isotopes to be observed, strengthens our interpretation that, also in these single isotope peak spectra true neuropeptide ions (albeit in very low quantities) are detected. This is also in line with the well-known very low abundance of typical signaling (secretory) peptides in tissues.

It follows as well that not all pixels may generate enough ion signals to allow for MS/MS identification/confirmation of the peptide primary structure. Yet when selecting areas with the highest MS signal for the respective precursor m/z, sufficient tandem MS information can be deduced from fragmentation spectra to allow for unambiguous annotation of the spectra to the respective neuropeptide. MALDI fragmentation spectra generated with the iMScope TRIO IT-TOF system are shown in **Figure 6**. These fully agree with the earlier described fragmentation spectra, when, in addition to the typical MS1 characteristics of typical endogenous peptide ions [16],

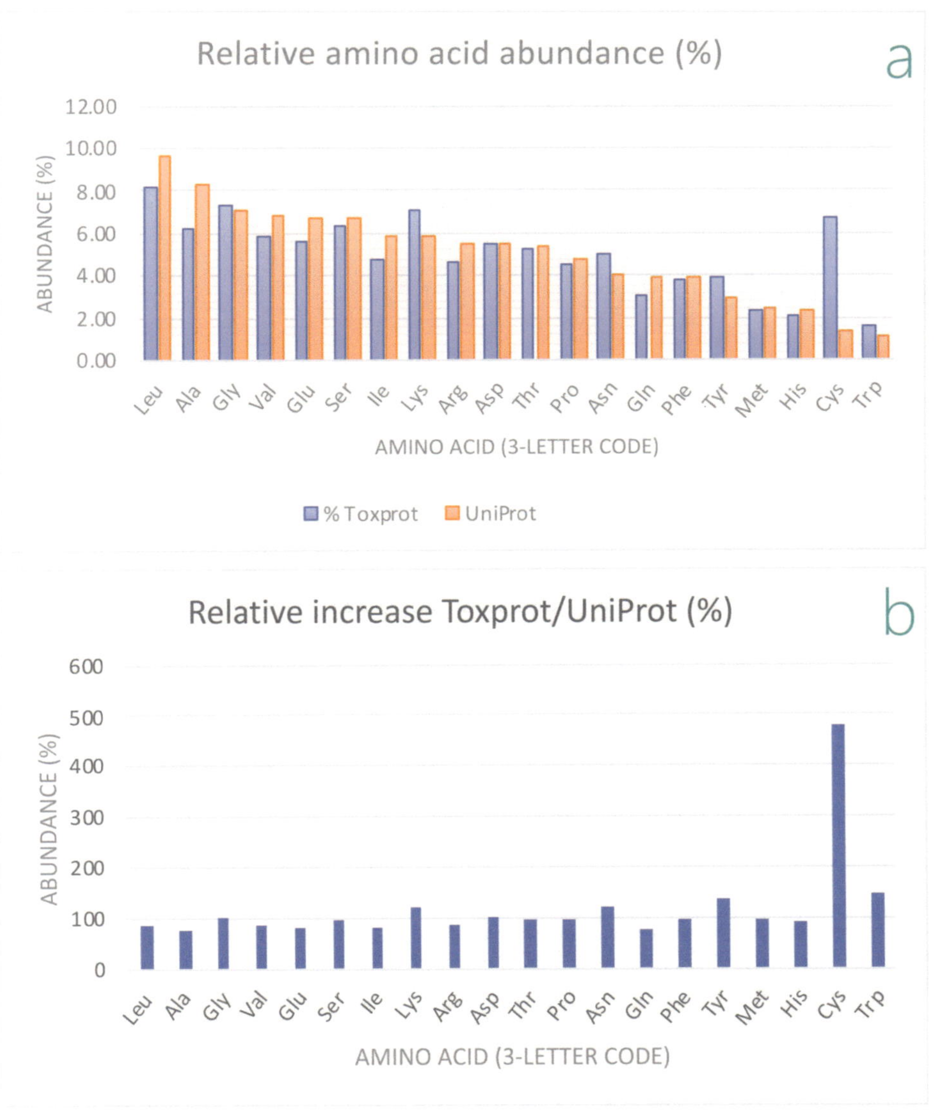

Figure 7.
Bar graphs illustrating importance of Cys as residue in secreted, bioactive proteins. (a) Upper panel compares relative amino acid abundance (in %, y-axis) in UniProt (red bars) with that in tox-Prot database (blue bars) as of January 2024 (a). Note especially the relatively high abundance of the cysteine residues in the Tox-Prot database (6.66%) compared to the Uniprot database (1.39%). (b) Lower panel shows an alternative representation of Cys importance by plotting relative amino acid abundance increases in Tox-Prot database compared to Uniprot database, calculated by (%Tox-Prot/%Uniprot) × 100%. Generally speaking, relative amino acid abundances are comparable between both databases, except for Cys which exhibits an almost 500% increase in relative abundance.

we also used ion mobility to confirm the identity of human pituitary nonapeptides extracted from a single FFPE tissue section [21].

Even using one of the earliest generation ion trap-orbitrap combinations (our LTQ Orbitrap Velos, first operated in 2010, being the oldest HRMS instrument tested), vasopressin as well as oxytocin ions can be detected [16, 17]. However, signal intensities seen with the later generation orbitraps [this chapter] are markedly higher, suggesting that, in the older-generation instruments, the low abundant endogenous

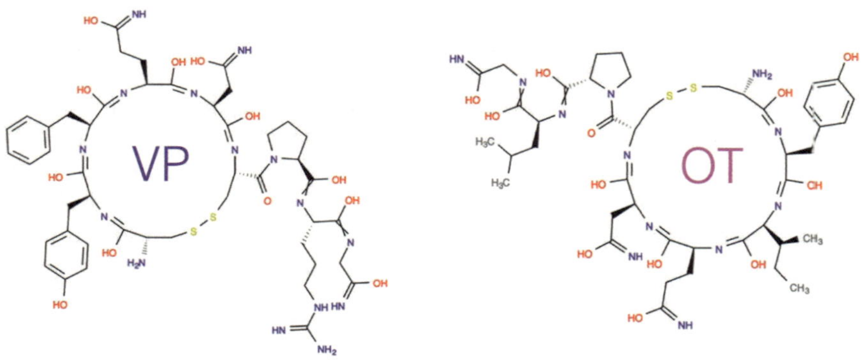

Figure 8.
Homo sapiens vasopressin (VP, left) and oxytocin (OI, right) in their predominant (proteo)forms, i.e., a carboxyterminally amidated nonapeptide with two cysteines at residues 1 and 6 linked by a disulfide bridge. Note: primary structure of vasopressin is CYFQNCPRG-NH_2 [elemental composition $C_{46}H_{65}N_{15}O_{12}S_2$], and of oxytocin is CYIQNCPLG-NH_2 [elemental composition $C_{43}H_{66}N_{12}O_{12}S_2$].

peptide analyses appear close to the limit of detection. In our view, this also explains why older-generation instruments may lack the sensitivity to reveal the complete isotopic envelope of a detected peptide. This then yields ambiguous data, which may not be conclusive, in particular when untargeted biomolecular discovery is envisaged, such as for novel biomarker identifications. We, therefore, recommend employing an HR mass spectrometer with the highest sensitivity possible for such analyses. We remark here that present-generation hybrid TOF instruments, such as the performant iMScope TRIO apparatus are also able to yield sufficient resolution as well as sensitivity to allow for MSHC detection/discovery of neuropeptides and/or neuropeptide proteoforms, even including on-tissue CID tandem MS confirmation of neuropeptide primary structures (**Figure 6**). This suggests that also its current successor, the iMScope QT, based on a QTOF mass spectrometer, may be well suited for this.

It is well-known that, in traversing the endoplasmic reticulum and Golgi apparatus, secretory proteins/peptides undergo a variety of posttranslational modifications (PTMs), with disulfide bond formation often being a PTM necessary for protein/peptide (secondary) structure and/or enhancing the peptide's half-life by hindering protease (such as trypsin) digestion. It follows that a relatively high proportion of bioactive secretory peptides share the attribute of having one or more Cys-residues in their primary structure. We have used this feature in the past to filter bioactive peptide candidates from complex LC/MS datasets of animal secretions [18]. The conventional method of determining the presence of disulfide bonds in a peptide is by simply counting the mass shift before and after reduction [22] or after reduction and alkylation [23]. However, such an approach requires the comparison of two separate analyses, which is not always feasible or desirable. To quickly assign the presence of cysteines in a peptide of unknown nature without the need for disulfide bond reduction, we, therefore, developed an HRMS filtering approach [18], in which we employ two intrinsic characteristics of the sulfur atom to select peptides with cysteine residues. The relatively large negative mass defect (the difference between the isotopic and nominal or integer mass) in combination with the positive isotopic shift (difference between average and monoisotopic mass) [24] allows for rapid identification of sulfur-containing peptide candidates from LCMS datasets of a complex mixture of unknown Cys-containing samples. Normalization of these two shifts results in two non-additive and independent peptide properties, which had previously been introduced as normalized nominal mass defect (NMD) and normalized isotopic shift (NIS), both of which have been used as a data representation method after de novo sequencing [25].

In the present chapter, we show that the detection and localization of Cys-containing secretory peptides can be done directly on the very same histological section as prepared by the histopathologist for H&E and/or IHC stains. The extremely high mass accuracy obtained by the various HR (Fourier transform, FT) MS instruments employed in these experiments allows for direct elemental composition calculation. The results generated are sufficient for database identification, such as those provided by the online automated mass spectrometry imaging data annotation platform METASPACE (https://metaspace2020.eu/ [26]).

Comparing the on-tissue isotope distributions of the neuropeptide-related ions with the theoretical calculations, we observe a-typical patterns. Indeed, in a significant number of pixels, the 3rd (and 4th) isotopic peak of vasopressin appears higher than expected (**Figure 2a**). We previously substantiated, using orthogonal validation with synthetic peptide analogs, tandem MS and ion mobility, that this is due to the presence of reduced neuropeptide, i.e. vasopressin with its [Cys1-Cys6] disulfide bridge opened [21]. In this chapter we show (single pixel) on-tissue MS spectra suggesting that a similar phenomenon is evident for oxytocin (**Figure 2c** and **d**). One biological assumption can be that we are observing the (in Golgi?) maturation of the neuropeptides, where in the process of peptide synthesis first the linear version of the peptide and its neurophysin are produced after which they are fully post-translationally processed, by C-terminal amidation and cystine formation. As it can be expected that the linear versions of both nonapeptides exhibit different binding characteristics to their respective neuropeptide receptors, the fact that we can now establish which neuropeptide proteoform is present in which tissue (compartment) may prove to be biologically relevant.

We also emphasize here that, although the tissues have been fixed in formaldehyde and stored in paraffin for many years (some were over a decade old), remarkably little formaldehyde fixation modifications are evident in our datasets. Only the pixels with the highest vasopressin concentration show signals at *m/z* 1096.44, being the Schiff base ion $[M + C + H]^+$ [particularly detected in the Exploris and Exactive series mass spectrometers; data not shown], a well-known intermediate reaction product of the formaldehyde fixation [27].

In general, we show here that MSHC results can be obtained on a wide variety of orbitrap systems (ThermoFisher Scientific), but also the fully integrated MS imaging IT-TOF instrument (Shimadzu). In combination with our previously published work [10, 21] which also employed Bruker mass spectrometers, this exemplifies that MSHC is full 'vendor-neutral', and that the unique sample preparation method is key to retaining the respective analytes of interest in their tissue location and thus allow for successful MSHC experiments.

Finally, we highlight that in terms of spatial resolution achievable with the different AP MALDI sources employed and using the very MSHC sample preparation described here, it was possible to generate human neuropeptide ion images from pixels as small as $5 \times 5\ \mu m^2$ (of a 5 μm thin FFPE tissue section).

5. Conclusions

We conclusively demonstrate that MSHC allows for different neuropeptides to be imaged on 5 μm thick histological sections of year-long biobanked FFPE tissue.

MSHC in addition allows for the visualization of the tissue localization of certain classes of metabolites (such as protoporphyrin and related biomolecules) as well. (These will be extensively reported elsewhere). The MSHC analyses yield sufficient resolution to show the on-tissue presence of both oxidized (cystine-containing) and reduced (separate cysteine residues) neuropeptides.

Whereas FFPE tissue is traditionally not considered, let alone recommended for MSI applications in pathology, using this combination of MSHC sample preparation and atmospheric pressure MALDI analysis, we demonstrate the value and potential of analyzing easily accessible FFPE biobanked tissue. MSHC analyses, therefore, have great promise to help understand pathological processes involving peptides, which today can only be studied with microscopic histo(patho)logy. Hence we can paraphrase the conclusion of Ref. [28] that MSHC as described here adds to the promise towards integration of MSI in routine molecular pathology for clinical diagnostics.

Acknowledgements

We would like to express our gratitude to Dr. Kerstin Strupat (ThermoFisher Scientific (Bremen) GmbH) for her continued support in bringing MALDI imaging and orbitrap mass spectrometry together. Prof. Dr. Uwe Karst is gratefully acknowledged for providing us access to his iMScope TRIO at the University of Münster (Germany). We are indebted to Dr. Venkat Panchagnula (MassTech, US) and Dr. Charles C. Liu (ASPEC Technologies, China) for their continued support and enthusiasm. We would also like to thank Anatech (South Africa) for their technical assistance and support.

PV would like to dedicate this book chapter to the memory of Dr. Martin Robert Lloyd Paine, who sadly passed away due to colon cancer in the middle of the COVID pandemic in May 2021. Marty, you will never be forgotten!

Author details

Peter Verhaert[1,2,3*], Gilles Frache[4], Dhaka Bhandari[5,6], Luuk Van Oosten[1], Remco Crefcoeur[1], Bernhard Spengler[5], Marthe Verhaert[1,7], Aletta Millen[2], Sooraj Baijnath[2], Ann-Christin Niehoff[8] and Raf Sciot[3]

1 ProteoFormiX BV, Vorselaar, Belgium

2 Faculty of Health Sciences, WITS Integrated Molecular Physiology Research Initiative, WITS Health Consortium (PTY) Ltd, School of Physiology, University of The Witwatersrand, Johannesburg, South Africa

3 Leuven University Department of Cell and Tissue Imaging, Leuven, Belgium

4 Luxembourg Institute of Science and Technology (LIST), Belvaux, Luxembourg

5 Justus-Liebig-Universität and TransMIT GmbH, Giessen, Germany

6 Gandaki Province Academy of Science and Technology, Pokhara, Nepal

7 Vrije Universiteit Brussel, Universitair Ziekenhuis Brussel, Jette, Belgium

8 Shimadzu Europa GmbH, European Innovation Center, Duisburg, Germany

*Address all correspondence to: peter.verhaert@proteoformix.com

References

[1] De Loof A. How to deduce and teach the logical and unambiguous answer, namely L = $\sum$C, to "What is Life?" using the principles of communication? Communicative & Integrative Biology. 2015;**8**:e1059977. DOI: 10.1080/19420889.2015.1059977

[2] De Loof A. The mega-evolution of life depends on sender-receiver communication and problem-solving. Theoretical Biology Forum. 2022;**115**:99-117. DOI: 10.19272/202211402007

[3] Verhaert P, Geysen J, De Loof A, Vandesande F. Immunoreactive material resembling vertebrate neuropeptides and neurophysins in the brain, suboesophageal ganglion, corpus cardiacum and corpus allatum of the dictyopteran *Periplaneta americana* L. Cell and Tissue Research. 1984;**238**:55-59. DOI: 10.1007/BF00215144

[4] Verhaert P, Marivoet S, Vandesande F, De Loof A. Localization of CRF immunoreactivity in the central nervous system of three vertebrate and one insect species. Cell and Tissue Research. 1984;**238**:49-53. DOI: 10.1007/BF00215143

[5] Starratt AN. Proctolin, an insect neuropeptide. Trends in Neurosciences. 1979;**2**:15-17. DOI: 10.1016/0166-2236(79)90008-0

[6] Jörnvall H, Agerberth B, Zasloff M. Viktor Mutt: A giant in the field of bioactive peptides. In: Skulachev VP, Semenza G, editors. Comprehensive Biochemistry. Vol. 46. Amsterdam: Elsevier; 2008. pp. 397-416. DOI: 10.1016/S0069-8032(08)00006-5

[7] Köllisch GV, Lorenz MW, Kellner R, Verhaert PD, Hoffmann KH. Structure elucidation and biological activity of an unusual adipokinetic hormone from corpora cardiaca of the butterfly, *Vanessa cardui*. European Journal of Biochemistry. 2000;**267**:5502-5508. DOI: 10.1046/j.1432-1327.2000.01611.x

[8] Verhaert P, Uttenweiler-Joseph S, de Vries M, Loboda A, Ens W, Standing KG. Matrix-assisted laser desorption/ionization quadrupole time-of-flight mass spectrometry: An elegant tool for peptidomics. Proteomics. 2001;**1**:118-131. DOI: 10.1002/1615-9861(200101)1:1<118::AID-PROT118>3.0.CO;2-1

[9] Verhaert PD, Pinkse MW, Strupat K, Prieto-Conaway MC. Imaging of similar mass neuropeptides in neuronal tissue by enhanced resolution MALDI MS with an ion trap - orbitrap hybrid instrument. Methods in Molecular Biology. 2010;**656**:433-449. DOI: 10.1007/978-1-60761-746-4_25

[10] Paine MRL, Ellis SR, Maloney D, Heeren RMA, Verhaert PDEM. Digestion-free analysis of peptides from 30-year-old formalin-fixed, paraffin-embedded tissue by mass spectrometry imaging. Analytical Chemistry. 2018;**90**:9272-9280. DOI: 10.1021/acs.analchem.8b01838

[11] Lemaire R, Desmons A, Tabet JC, Day R, Salzet M, Fournier I. Direct analysis and MALDI imaging of formalin-fixed, paraffin-embedded tissue sections. Journal of Proteome Research. 2007;**6**:1295-1305. DOI: 10.1021/pr060549i

[12] Stauber J, Lemaire R, Franck J, Bonnel D, Croix D, Day R, et al. MALDI imaging of formalin-fixed paraffin-embedded tissues: Application to

model animals of Parkinson disease for biomarker hunting. Journal of Proteome Research. 2008;**7**:969-978. DOI: 10.1021/pr070464x

[13] Chaurand P, Latham JC, Lane KB, Mobley JA, Polosukhin VV, Wirth PS, et al. Imaging mass spectrometry of intact proteins from alcohol-preserved tissue specimens: Bypassing formalin fixation. Journal of Proteome Research. 2008;**7**:3543-3555. DOI: 10.1021/pr800286z

[14] Kröger S, Niehoff AC, Jeibmann A, Sperling M, Paulus W, Stummer W, et al. Complementary molecular and elemental mass-spectrometric imaging of human brain tumors resected by fluorescence-guided surgery. Analytical Chemistry. 2018;**90**:12253-12260. DOI: 10.1021/acs.analchem.8b03516

[15] Römpp A, Schramm T, Hester A, Klinkert I, Both JP, Heeren RM, et al. imzML: Imaging mass spectrometry markup language: A common data format for mass spectrometry imaging. Methods in Molecular Biology. 2011;**696**:205-224. DOI: 10.1007/978-1-60761-987-1_12

[16] Bhattacharya N, Nagornov K, Verheggen K, Verhaert M, Sciot R, Verhaert P. MS1-based data analysis approaches for FFPE tissue imaging of endogenous peptide ions by mass spectrometry histochemistry (MSHC). Methods in Molecular Biology. 2023;**2688**:187-202. DOI: 10.1007/978-1-0716-3319-9_16

[17] Verheggen K, Bhattacharya N, Verhaert M, Goossens B, Sciot R, Verhaert P. HistoSnap: A novel software tool to extract *m/z*-specific images from large MSHC datasets. Methods in Molecular Biology. 2023;**2688**:15-26. DOI: 10.1007/978-1-0716-3319-9_2

[18] van Oosten LN, Pieterse M, Pinkse MWH, Verhaert PDEM. Screening method for the discovery of potential bioactive cysteine-containing peptides using 3D mass mapping. Journal of the American Society for Mass Spectrometry. 2015;**26**:2039-2050. DOI: 10.1007/s13361-015-1282-z

[19] Verhaert PD, Sciot R, Frache G. Pushing the limits of atmospheric pressure MALDI MS histochemistry on high-resolution OrbitrapTM systems for FFPE patient tissue analysis. In: Proceedings of the 71st Conference on Mass Spectrometry and Allied Topcis (ASMS2023; June 4-8, 2023) Houston, TX. Houston: American Society for Mass Spectrometry (ASMS); 2023. p. MP158

[20] McDonnell LA, Piersma SR, Altelaar MAF, Mize TH, Luxembourg SL, Verhaert PD, et al. Subcellular imaging mass spectrometry of brain tissue. Journal of Mass Spectrometry. 2005;**40**:160-168. DOI: 10.1002/jms.735

[21] Cintron-Diaz YL, Gomez-Hernandez ME, Verhaert MMHA, Verhaert PDEM, Fernandez-Lima F. Spatially resolved neuropeptide characterization from neuropathological formalin-fixed, paraffin-embedded tissue sections by a combination of imaging MALDI FT-ICR mass spectrometry histochemistry and liquid extraction surface analysis-trapped ion mobility spectrometry-tandem mass spectrometry. Journal of the American Society for Mass Spectrometry. 2022;**33**:681-687. DOI: 10.1021/jasms.1c00376

[22] Quinton L, Demeure K, Dobson R, Gilles N, Gabelica V, De Pauw E. New method for characterizing highly disulfide-bridged peptides in complex mixtures: Application to toxin identification from crude venoms. Journal of Proteome Research. 2007;**6**:3216-3223. DOI: 10.1021/pr070142t

[23] Neitz S, Jürgens M, Kellmann M, Schulz-Knappe P, Schrader M. Screening for disulfide-rich peptides in biological sources by carboxyamidomethylation in combination with differential matrix-assisted laser desorption/ionization time-of-flight mass spectrometry. Rapid Communications in Mass Spectrometry. 2001;**15**:1586-1592. DOI: 10.1002/rcm.413

[24] Sleno L. The use of mass defect in modern mass spectrometry. Journal of Mass Spectrometry. 2012;**47**(2):226-236. DOI: 10.1002/jms.2953

[25] Artemenko KA, Zubarev AR, Samgina TY, Lebedev AT, Savitski MM, Zubarev RA. Two dimensional mass mapping as a general method of data representation in comprehensive analysis of complex molecular mixtures. Analytical Chemistry. 2009;**81**:3738-3745. DOI: 10.1021/ac802532j

[26] Palmer A, Phapale P, Chernyavsky I, Lavigne R, Fay D, Tarasov A, et al. FDR-controlled metabolite annotation for high-resolution imaging mass spectrometry. Nature Methods. 2017;**14**:57-60. DOI: 10.1038/nmeth.4072

[27] Klockenbusch C, O'Hara JE, Kast J. Advancing formaldehyde cross-linking towards quantitative proteomic applications. Analytical and Bioanalytical Chemistry. 2012;**404**:1057-1067. DOI: 10.1007/s00216-012-6065-9

[28] Huizing LRS, Ellis SR, Beulen BWAMM, Barré FPY, Kwant PB, Vreeken RJ, et al. Development and evaluation of matrix application techniques for high throughput mass spectrometry imaging of tissues in the clinic. Clinical Mass Spectrometry. 2019;**12**:7-15. DOI: 10.1016/j.clinms.2019.01.004

Chapter 3

The Relationship between Cysteine, Homocysteine, and Osteoporosis

Alexandru Filip, Bogdan Veliceasa, Gabriela Bordeianu, Cristina Iancu, Magdalena Cuciureanu and Oana Viola Badulescu

Abstract

Both cysteine and homocysteine are sulfur-containing amino acids that play distinct roles in the body. Cysteine is an amino acid that contributes to the synthesis of collagen, a crucial protein for bone structure. Collagen provides the structural framework for bones, contributing to their strength and flexibility. Adequate collagen formation is vital for maintaining bone integrity, and cysteine's role in collagen synthesis suggests a potential indirect impact on bone health. Elevated levels of homocysteine have been associated with an increased risk of osteoporosis and bone fractures. The exact mechanisms through which homocysteine affects bone metabolism are not fully understood, but it is suggested to involve interference with collagen cross-linking, increased oxidative stress, and altered bone remodeling. The relationship between cysteine, homocysteine, and osteoporosis is intertwined within complex biochemical pathways, constituting a continually evolving area of research.

Keywords: osteoporosis, bone mineral density, elderly, cysteine, homocysteine

1. Introduction

Both cysteine (Cys) and homocysteine (Hcy) are amino acids that contain a sulfur atom. The amino acid cysteine can be obtained from both the diet and produced from the degradation of methionine through the transsulfuration pathway [1, 2]. Methionine, another sulfur-containing essential amino acid, is found in a variety of protein-rich foods, including meat, eggs, dairy products, and legumes [3]. Homocysteine is a sulfur-containing amino acid that is not directly obtained from dietary protein or utilized for its endogenous synthesis. Instead, it functions as an intermediate in the metabolic pathway of methionine. Elevated plasma levels of homocysteine, known as hyperhomocysteinemia, are a risk factor for various pathological disorders. Homocysteine metabolism is influenced by B vitamins, and there have been suggestions for vitamin treatments to lower Hcy concentration [4, 5]. However, clinical trials have not yet established a consensus on the effectiveness of vitamin supplements in reducing homocysteine levels and improving pathological conditions [6]. This is particularly evident in elderly patients with associated pathologies, indicating that these dietary, as well as non-dietary factors may contribute to elevated homocysteine levels [6].

A frequently encountered condition, especially in the elderly, is osteoporosis with serious health consequences due to its association with fragility fractures. As a result of the increase in life expectancy, the total number of fragility fractures will increase and, therefore, the cost to society will increase significantly in the coming years. Identifying risk factors or risk indicators to prevent osteoporosis and develop new treatment strategies are real concerns [7].

The link between homocysteine and the bone system has been observed in studies of hyperhomocysteinuria. People with hyperhomocysteinuria have numerous skeletal defects, including reduced BMD and osteopenia [8]. Elevated levels of homocysteine have been correlated with osteoporosis and an increased incidence of hip fractures in postmenopausal women. The increased bone fragility in this situation has been explained as a result of the change in the bone matrix when homocysteine interferes with collagen cross-linking [9]. Many studies report that Hcy and vitamins involved in its metabolism, such as folic acid and vitamins B6 and B12, affect bone metabolism, bone quality, and fracture risk, especially in the elderly. Considering the previously mentioned Hcy could be seen as a risk indicator for osteoporosis related to micronutrient deficiency [10–15]. Experimental results indicate that cysteine (Cys) is involved in bone metabolism by incorporation into collagen and cysteine protease enzymes [16]. The association between osteoporosis and the decrease in cysteine was explained either by the reduced availability of cysteine for collagen formation or due to the increased use of cysteine by proteases in the osteoclastic hyperactivity that underlies the osteoporotic process [17]. Fractures in osteoporosis represent a severe manifestation of the condition, often accompanied by visible clinical symptoms. Collagen stands out as a vital element in the bone matrix, and any disruption of its cross-linking formation increases susceptibility to fracture. Vitamin B6 serves as a key regulator in cross-linked collagen formation. Vitamin B6 deficiency correlates with reduced bone strength due to its impact on collagen reticular disruption, both directly and *via* homocysteine-related pathways.

2. Overview of cysteine and homocysteine metabolism

Cysteine is primarily synthesized through the transsulfuration pathway, which involves the conversion of homocysteine to cysteine. Cysteine biosynthesis uses the sulfur group of methionine with the carbon backbone of serine. This pathway intersects with the methionine cycle, which supports transmethylation reactions and forms homocysteine as a key intermediate. This process occurs primarily in the liver and involves several enzymatic steps, including the action of cystathionine beta-synthase (CBS) and cystathionine gamma-lyase (CGL) (**Figure 1**) [18].

The transsulfuration pathway is regulated by various factors, including enzyme activity, substrate availability, and the presence of cofactors, such as vitamin B6 (pyridoxal phosphate). In addition to de novo synthesis, cysteine can also be obtained from dietary sources rich in sulfur-containing amino acids, such as methionine.

The first reaction of methionine metabolism is its transformation into S-adenosylmethionine (SAM), a process catalyzed by the enzyme methionine adenosyltransferase (MAT). SAM is transformed into S-adenosylhomocysteine (SAH). The product of this reaction, S-adenosylhomocysteine, is hydrolyzed by S-adenosylhomocysteine hydrolase to yield adenosine and homocysteine [19]. Homocysteine is combined with serine through the action of cystathionine β-synthase

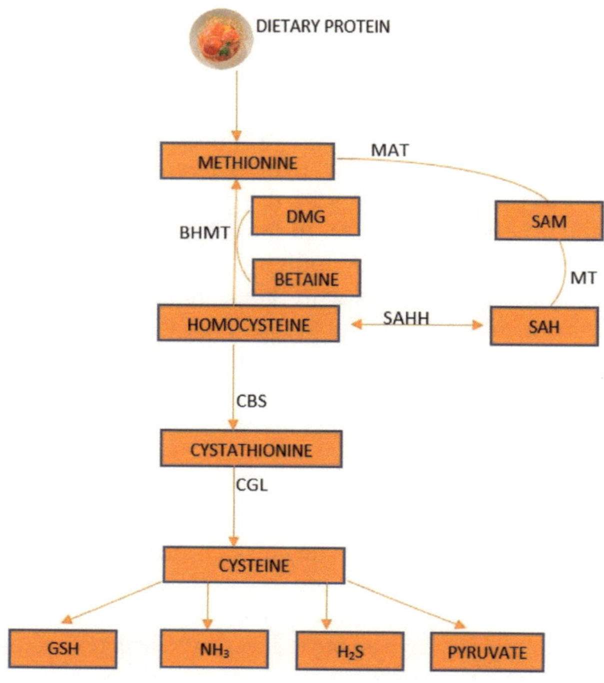

Figure 1.
Schematic cysteine and homocysteine pathways. Methionine adenosyltransferase (MAT); methyltransferase (MT); S-adenosylhomocysteinase hydrolase (SAHH); betaine-homocysteine methyltransferase (BHMT); cystathionine β-synthase (CBS); cystathionine gamma-lyase (CGL); glutathione (GSH); hydrogen sulfide (H2S). Adapted from Al-Sadeq et al. [18].

(CBS), resulting in the formation of cystathionine. Subsequently, cystathionine gamma-lyase catalyzes the cleavage of cystathionine, yielding cysteine. The primary role of CGL is to catalyze the cleavage of cystathionine, resulting in the production of cysteine, α-ketobutyrate, and ammonia, which occurs in the second step of the transsulfuration pathway. Additionally, CGL can catalyze side reactions with either cysteine, homocysteine, or both as substrates, potentially leading to the production of hydrogen sulfide [20]. In the process of transsulfuration, serine contributes its carbon chain to cysteine, while the sulfur atom required for cysteine synthesis originates from methionine [21]. Cysteine can be further converted to glutathione, ammonia, pyruvate, and hydrogen sulfide [18, 22]. The synthesis of cysteine through the transsulfuration pathway can be seen as a component of the degradation process of methionine and homocysteine. Cysteine acts as a carrier for the conversion of sulfur from methionine and homocysteine into end products that can be eliminated through urine. Enzymes involved in these pathways are subject to regulation by cofactors such as vitamin B6, vitamin B12, and folate, among others. Deficiencies in these cofactors can lead to disruptions in cysteine and homocysteine metabolism, potentially resulting in health issues, such as cardiovascular diseases and neural tube defects [23–25]. Cysteine and homocysteine metabolism intersect with several other metabolic pathways. For instance, cysteine serves as a precursor for the synthesis of glutathione, an important antioxidant [26]. Additionally, homocysteine metabolism is interconnected with the folate and methionine cycles, which are essential for DNA synthesis and methylation reactions [27].

3. Osteoporosis

Osteoporosis is a condition characterized by reduced bone mass and microarchitectural deterioration of bone tissue. The consequence is increased bone fragility and increased susceptibility to fractures [28]. Although this condition is commonly associated with the elderly, it can also affect young people (juvenile osteoporosis) or adults who are undergoing treatment with drugs such as steroids, chemotherapy, anticonvulsants, and others [29]. Changes in the mechanical properties of bones due to aging are noteworthy. Comparing individuals at an advanced age with those at 30 years of age, there is an 8% decrease in elastic modulus in cortical bone, a 64% decrease in trabecular bone, an 11% reduction in peak strength in cortical bone, and a substantial decrease with 68% in trabecular bone. In addition, hardness decreases by 34% in cortical bone and by 70% in trabecular bone [30].

3.1 Bone remodeling

Bone, a fundamental element of the human skeleton, is characterized by its robust, solid, and durable composition. They play crucial physiological roles, providing mechanical support, protecting vital organs, facilitating hematopoiesis, and contributing to mineral homeostasis. Bones can be classified according to position, shape, size, or structure. Bones have an outer layer called cortex, smooth, compact, continuous, and of variable thickness, and inside the bone tissue is arranged in a network of intersecting plates and spicules called trabeculae [31, 32].

In developed bone, the trabeculae are organized systematically, forming uninterrupted sections of bone tissue running parallel to the directions of compressive or tensile stress. These trabeculae create a sophisticated network of internal cross-shaped rods designed to enhance the bone's stiffness [33]. Bones exposed primarily to compressive or tensile forces typically possess thin cortices and achieve structural strength *via* trabeculae. In contrast, bones. Such as the femur, which endure bending, shearing, or torsional forces, often feature thick cortices, a tubular structure, and a continuous central cavity known as the medullary cavity. Bone tissue is made up of an organic matrix called osteoid, made up of collagen fibers and proteoglycans, which are impregnated with salts, mainly made up of calcium and phosphate; the main crystallized salt is hydroxyapatite [34]. Bone tissue undergoes periodic remodeling; under physiological conditions, the rate of bone tissue formation is in balance with the rate of bone resorption. Bone diseases in the elderly are associated with increased morbidity and mortality. Osteoporosis is a disease that affects the quality of life in the elderly [35]. Although bone growth stops after adolescence, bone is a very dynamic tissue. Bone tissue is continuously resorbed and regenerated through constant structural change. One of the important characteristics of bones is the remodeling capacity. This process of bone remodeling occurs during growth and continues throughout life [36]. Bone remodeling occurs continuously, resulting in the arrangement of bone structure into regular units, enhancing its ability to withstand mechanical forces effectively. Osteoclasts, responsible for bone resorption, remove older bone, while osteoblasts, involved in bone formation, initiate the process of new bone formation. Osteoclasts secrete proteolytic lysosomal enzymes, which efficiently break down the bone matrix akin to acids, converting calcium salts into a soluble form that can be absorbed into the bloodstream [37]. The remodeling process holds particular significance for the long bones responsible for supporting the limbs. These bones feature wider ends and narrower middles, imparting additional strength to the joints. Osteoclasts play a vital role within the bones, operating within the bone marrow cavity and cancellous bone spaces to widen these areas. Additionally, they work on the outer surfaces, diminishing bony processes like the epiphyseal regions of the upper and lower limbs [32]. Osteoclasts are active behind the epiphyseal growth zone. Their function within the bone involves breaking down immature bone, facilitating its transformation into compact, mature bone (known as lamellar bone). This process entails the release of elongated tubular spaces, which act as nuclei for the formation of osteons. Osteons are the bony structures through which blood vessels pass, contributing to the bone's vascular network and overall integrity [38]. While osteoclasts degrade old bone at the epithelial ends of the bone, osteoblasts within the growth zone initiate the formation of a new epiphysis. Concurrently, within each tubular space liberated by osteoclasts inside the bone, osteoblasts play a crucial role in depositing new layers of bone, thus contributing to bone remodeling and maintenance [39].

A bone that remains unused, such as a lower limb immobilized after trauma, is susceptible to resorption. This is because the rate of bone destruction surpasses the rate of bone formation in such circumstances [40].

Bones experiencing heightened stress undergo permanent remodeling. For instance, the femur undergoes complete replacement approximately every 6 months. Beyond altering bone structure, remodeling also plays a role in regulating the concentration of ionic calcium in the bloodstream. This mineral is essential for various physiological processes such as nerve signaling, cell membrane integrity, and blood clotting [41]. Bone turnover is increased in women after menopause because the rate of bone resorption is greater than that of bone formation [42].

3.2 Fragility fractures

The primary consequence associated with osteoporosis is the heightened susceptibility to fractures. While there exists a recognized correlation between bone density and the risk of osteoporotic fractures, it is not inevitable that every individual diagnosed with osteoporosis will experience fractures [43]. The likelihood of fractures occurring is contingent not only upon bone fragility but also on the level of trauma sustained. Typically, osteoporotic fractures are linked to falls from low heights, a scenario for which the elderly are particularly prone [44, 45]. Various factors contribute to the heightened propensity for falls among the elderly, including diminished visual acuity, vestibular dysfunction, cognitive impairment, musculoskeletal conditions, and the usage of certain medications [46–48].

Nonetheless, severe hypotension frequently emerges as a predominant characteristic. In cases where osteoporosis is suspected, assessing bone mineral density stands out as the premier diagnostic approach, aiding physicians in gauging fracture susceptibility. Fracture risk assessment can be performed using both clinical methods, which entail identifying risk factors, and paraclinical approaches, which encompass various techniques for evaluating bone properties. Individuals diagnosed with fragility fractures demonstrate an increased relative risk (RR) of suffering subsequent fractures. Patients diagnosed with fragility fractures exhibit an elevated relative risk (RR) of experiencing subsequent fractures. The principal risk factors associated with fractures (RR $\geq$ 2) include age, bone mineral density, previous history of fragility fractures, family history of fragility fractures, early menopause before the age of 45, glucocorticoid therapy, extended periods of immobilization, susceptibility to falls, malabsorption disorders, chronic renal failure, and undergoing transplantation [49–54].

Moderate risk factors (where the relative risk, RR, falls between 1 and 2) include conditions such as rheumatoid arthritis, Bechterew's disease, and the use of anticonvulsant medications. Additionally, low calcium intake, diabetes insipidus, estrogen deficiency, primary hyperparathyroidism, hyperthyroidism, smoking, and excessive alcohol consumption are also considered moderate risk factors for certain health conditions [49, 55].

There is a clear relationship between these risk factors and low bone density or other causes of osteoporotic fractures. For the correct identification of possible risk factors, both the anamnesis and the physical examination of the patients must be completed with biochemical tests (blood count, ESR, calcium, kidney and liver tests, alkaline phosphatase) and markers of bone turnover [56–58]. Obesity has been considered advantageous for maintaining healthy bones due to the higher bone mineral density seen in overweight individuals [59]. More than half of women and approximately one-third of men will experience osteoporotic fractures at some point in their lives. While there may be no symptoms prior to the fracture, assessing bone mineral density and other risk factors can help identify individuals at high risk. It is crucial to prioritize the measurement of bone mineral density and identification of other risk factors to accurately gauge the likelihood of fracture in patients, whether or not they have a history of previous fractures. This is imperative because fractures, often devoid of symptoms other than the fracture itself, can lead to severe consequences and may signal disease progression. The impact of obesity on fracture risk varies depending on the location, leading to increased risk for certain fractures, such as those involving the humerus or ankle, while decreasing the risk for others, such as fractures of the hip or pelvis [60].

Over the past two decades, more advanced technology has been developed for determining bone mass, and more techniques are available. With these bone densitometry techniques, the clinician can detect low bone mass prior to fracture. This is beneficial for patients because early treatment of osteoporosis can be initiated and thus osteoporotic fractures can be prevented.

In addition to the usual medical history considerations for trauma patients, special attention must be given to specific issues in elderly individuals. Assessing pre-injury mobility, concurrent medical conditions and medications, as well as cognitive and nutritional status is crucial. These factors significantly influence decision-making and the planning of therapy in collaboration with the patient.

4. Involvement of cysteine and homocysteine in osteoporosis

Four mechanisms are proposed in the literature regarding the involvement of homocysteine in bone remodeling: increased osteoclasts activity, decreased osteoblasts activity, decreased bone blood flow, and direct action of the Hcy molecule on the bone matrix [7].

Several studies regarding the link between homocysteine and bones are presented in **Table 1**.

Numerous studies indicate that elevated plasma levels of homocysteine are associated with osteoporotic fractures [62–64, 71]. The exact mechanism proposed to explain this association is not elucidated. One proposed hypothesis suggests that hyperhomocysteinemia could impact the bone matrix and reduce bone quality by hindering the formation of cross-links between collagen fibers, thus impeding the development of a robust collagen network structure [72]. It is assumed that Hcy interferes with collagen cross-linking in bone, thereby weakening the bone structure [72]. Another hypothesis

Study ID	Outcomes	Conclusion
Miyao et al. [61]	Polymorphism of methylenetetrahydrofolate reductase (MTHFR), bone mineral density	The MTHFR gene polymorphism demonstrates a notable correlation with bone mineral density (BMD) in postmenopausal women, affecting both lumbar and total body BMD levels
Van Meurs et al. [62]	Homocysteine levels and the risk of incident osteoporotic fracture	Elevated homocysteine levels emerge as a robust and independent risk factor for osteoporotic fractures in older individuals, encompassing both men and women
Gjesdal et al. [63]	Hcy, folate, and vitamin B12 and the methylenetetrahydrofolate reductase (MTHFR) 677C → T and 1298A → C polymorphisms	High total homocysteine levels and low folate levels were linked to decreased bone mineral density in women, whereas no such association was observed in men
Perier et al. [64]	Homocysteine, serum albumin, and serum creatinine	Plasma homocysteine levels exhibit a significant correlation with aging and serve as a marker of frailty, indicative of overall health and nutritional status, physical activity, and renal function impairment
Rumbak et al. [65]	Homocysteine, serum folate, red blood cell folate, and serum vitamin B12	Levels of homocysteine, folate, or vitamin B12 showed no association with bone mineral density (BMD) in a population of healthy Croatian women aged 45–65
Jianbo et al. [66]	Plasma total homocysteine, glucose, serum creatinine, plasma insulin, serum osteocalcin, serum 25-hydroxyvitamin D, serum folate, and vitamin B12	Diabetic patients with osteoporosis exhibited higher levels of homocysteine compared to those without osteoporosis
Sandeep et al. [67]	Serum homocysteine levels	Individuals with elevated circulating levels of homocysteine demonstrated a decreased bone mineral density (BMD), establishing an association between homocysteine and the risk of developing osteoporosis
Garcia-Alfaro et al. [68]	Hcy, creatinine, calcium, phosphorus, vitamin D, and parathormone levels	Plasma levels of homocysteine do not show a relationship with BMD in the lumbar spine (L1–L4), femoral neck, or total hip
Widjaja et al. [69]	Homocysteine	Elevated homocysteine levels did not increase the incidence of osteoporotic fractures. However, homocysteine levels increased with aging and correlated with bone mineral density
Watanabe et al. [70]	Total procollagen type 1 amino-terminal propeptide (total P1-NP), intact parathyroid hormone (intact PTH), and homocysteine	Preoperative osteoporosis and elevated serum homocysteine levels were identified as risk factors for intraoperative periprosthetic fractures

Table 1.
Research on the relationship between homocysteine and bone.

suggests that hyperhomocysteinemia triggers the formation and activity of osteoclasts by generating intracellular reactive oxygen species (ROS) within the bone marrow, thereby elevating bone resorption. Moreover, heightened levels of homocysteine have been associated with reduced expression of osteocalcin and increased expression of osteopontin, disrupting normal osteoblast function. This disruption ultimately leads to diminished bone formation and contributes to the development of osteoporosis [72, 73].

Another hypothesis proposed refers to the fact that homocysteine, involved in the methylation process, could disrupt DNA methylation and gene expression, which can lead to changes in bone structure. The hypothesis suggesting that changes in methylation capacity could influence bone metabolism is bolstered by research demonstrating a correlation between a decreased S-adenosylmethionine (SAM) to S-adenosylhomocysteine (SAH) ratio in bone and diminished bone strength observed in hyperhomocysteinemic rats [71]. The results of the study conducted by Garcia-Alfaro et al. [68] evaluated a group of postmenopausal women and reported that homocysteine levels were not related to BMD in the lumbar spine (L1-L4), femoral neck, and total hip [68].

Conflicting data persists regarding the relationship between high homocysteine levels and low bone mineral density. Epidemiological evidence is supported by genetic studies showing an association between the common MTHFR 677C_T polymorphism and osteoporosis risk.

The concentrations of cystine and cysteine were found to be reduced in individuals with osteoporosis, suggesting that cysteine may act as a protective factor for bone mineral density [74, 75]. Cystine and cysteine may accelerate bone regeneration by activating osteoblast (OB) differentiation [76]. Bone regeneration frequently relies on signals from osteogenesis-inducing factors for a successful outcome. N-acetyl cysteine (NAC), a small antioxidant molecule, potentially regulates osteoblastic differentiation [76]. The study by Yamada et al. [76] indicates that N-acetyl cysteine (NAC) can serve as an osteogenesis-enhancing molecule, accelerating bone regeneration by stimulating the differentiation of osteogenic lineages. The systematic review and meta-analysis of different metabolites in osteopenia and osteoporosis by Wang et al. [77] indicated a positive association between cystine and BMD.

Baines et al. [16] examined the relationship between plasma cysteine and related thiols, as well as bone turnover markers, and folate and vitamin B6 levels with calcaneal bone mineral density (BMD) in 328 women, categorized based on their BMD measurements [16]. Their findings revealed a noteworthy association between plasma Cys levels and bone mineral density. A decrease in Cys concentration, potentially influenced by smoking or reduced homocysteine flux, could diminish its availability for collagen formation [16]. This might prompt heightened osteoclast activation, potentially due to relative hyperhomocysteinemia, leading to increased utilization of Cys in cysteine proteases.

5. Conclusions

It is evident that changes in the metabolism of cysteine and homocysteine can contribute to a disorder of bone remodeling, favoring the occurrence of osteoporosis. Increased Hcy levels result in the dysregulation of the transsulfuration pathway as L-cysteine is synthesized from Hcy. Epidemiological cohort studies demonstrate robust associations between serum homocysteine concentrations and a heightened incidence of fractures. The results of clinical trials investigating vitamin treatments for Hcy lowering have not established a consensus on the effectiveness of vitamin supplementation in reducing homocysteine levels, and thereby osteoporosis.

Conflict of interest

The authors declare no conflict of interest.

Author details

Alexandru Filip[1*], Bogdan Veliceasa[1], Gabriela Bordeianu[2], Cristina Iancu[3], Magdalena Cuciureanu[4] and Oana Viola Badulescu[5]

1 Faculty of Medicine, Department of Orthopedics and Traumatology, Surgical Science II, "Grigore T Popa" University of Medicine and Pharmacy, Iasi, Romania

2 Faculty of Medicine, Department of Morpho-Functional Sciences II, Discipline of Biochemistry, "Grigore T Popa" University of Medicine and Pharmacy, Iasi, Romania

3 Faculty of Pharmacy, Discipline of Biochemistry, "Grigore T Popa" University of Medicine and Pharmacy, Iasi, Romania

4 Faculty of Medicine, Department of Morpho-Functional Sciences II, Discipline of Pharmacology, "Grigore T. Popa" University of Medicine and Pharmacy, Iasi, Romania

5 Faculty of Medicine, Department of Morpho-Functional Sciences II, Discipline of Pathophysiology, "Grigore T. Popa" University of Medicine and Pharmacy, Iasi, Romania

*Address all correspondence to: filipzamosteanu@gmail.com

References

[1] Kohl JB, Mellis AT, Schwarz G. Homeostatic impact of sulfite and hydrogen sulfide on cysteine catabolism. British Journal of Pharmacology. 2019;**176**(4):554-570. DOI: 10.1111/bph.14464. Epub 2018 Sep 27

[2] Kobayashi S, Lee J, Takao T, Fujii J. Increased ophthalmic acid production is supported by amino acid catabolism under fasting conditions in mice. Biochemical and Biophysical Research Communications. 2017;**491**(3):649-655. DOI: 10.1016/j.bbrc.2017.07.149. Epub 2017 Jul 28

[3] Wu DF, Yin RX, Deng JL. Homocysteine, hyperhomocysteinemia and H-type hypertension. European Journal of Preventive Cardiology. 2024;**31**:zwae022. DOI: 10.1093/eurjpc/zwae022

[4] Zhang S, Lv Y, Luo X, Weng X, Qi J, Bai X, et al. Homocysteine promotes atherosclerosis through macrophage pyroptosis via endoplasmic reticulum stress and calcium disorder. Molecular Medicine. 2023;**29**(1):73. DOI: 10.1186/s10020-023-00656-z

[5] Monzani D, Liberale C, Segato E, De Cecco F, Arietti V, Palma S, et al. The role of fibrinogen, homocysteine and metabolic syndrome's alterations in sudden sensorineural hearing loss (SSHL): A narrative review. Medicina (Kaunas, Lithuania). 2023;**59**(11):1977. DOI: 10.3390/medicina59111977

[6] Azzini E, Ruggeri S, Polito A. Homocysteine: Its possible emerging role in at-risk population groups. International Journal of Molecular Sciences. 2020;**21**(4):1421. DOI: 10.3390/ijms21041421

[7] Herrmann M, Widmann T, Herrmann W. Homocysteine—A newly recognised risk factor for osteoporosis. Clinical Chemistry and Laboratory Medicine. 2005;**43**(10):1111-1117. DOI: 10.1515/CCLM.2005.194

[8] Fratoni V, Brandi ML. B vitamins, homocysteine and bone health. Nutrients. 2015;7:2176-2192. DOI: 10.3390/nu7042176

[9] Martí-Carvajal AJ, Solà I, Lathyris D, Dayer M. Homocysteine-lowering interventions for preventing cardiovascular events. Cochrane Database of Systematic Reviews. 2017;**8**(8):CD006612. DOI: 10.1002/14651858.CD006612.pub5

[10] Swart KMA, van Schoor NM, Lips P. Vitamin B12, folic acid, and bone. Current Osteoporosis Reports. 2013;**11**:213-218

[11] Salari P, Abdollahi M, Heshmat R, Meybodi HA, Razi F. Effect of folic acid on bone metabolism: A randomized double-blind clinical trial in postmenopausal osteoporotic women. DARU Journal of Pharmaceutical Sciences. 2014;**22**:1-7

[12] Halıloglu B, Aksungar FB, Ilter E, Peker H, Akın FT, Ozekıcı U. Relationship between bone mineral density, bone turnover markers and homocysteine, folate and vitamin B12 levels in postmenopausal women. Archives of Gynecology and Obstetrics. 2010;**281**:663-668

[13] Stone KL, Lui LY, Christen WG, Troen AM, Bauer DC, Kado D, et al. Effect of combination folic acid, vitamin B6, and vitamin B12 supplementation on fracture risk in women: A randomized, controlled trial. Journal of Bone and Mineral Research. 2017;**32**(12):2331-2338

[14] Haliloglu B, Peker H. Postmenopausal homocysteine, vitamin B12, folate levels and bone metabolism: A focus on fractures. Nutrition and Diet in Menopause. 2013:91-99

[15] Bozkurt N, Erdem M, Yılmaz E, Erdem A, Biri A, Kubatova A, et al. The relationship of homocyteine, B12 and folic acid with the bone mineral density of the femur and lumbar spine in Turkish postmenopausal women. Archives of Gynecology and Obstetrics. 2009;**280**:381-387

[16] Baines M, Kredan MB, Davison A, et al. The association between cysteine, bone turnover, and low bone mass. Calcified Tissue International. 2007;**81**:450-454. DOI: 10.1007/s00223-007-9089-y

[17] Elshorbagy AK, Gjesdal CG, Nurk E, Tell GS, Ueland PM, Nygård O, et al. Cysteine, homocysteine and bone mineral density: A role for body composition? Bone. 2009;**44**(5):954-958. DOI: 10.1016/j.bone.2008.12.018

[18] Al-Sadeq DW, Nasrallah GK. The spectrum of mutations of homocystinuria in the MENA region. Genes. 2020;**11**:330. DOI: 10.3390/genes11030330

[19] Carballal S, Banerjee R. Overview of cysteine metabolism. In: Redox Chemistry and Biology of Thiols. Cambridge, Massachusetts: Academic Press; 2022. pp. 423-450

[20] Stipanuk MH, Ueki I. Dealing with methionine/homocysteine sulfur: Cysteine metabolism to taurine and inorganic sulfur. Journal of Inherited Metabolic Disease. 2011;**34**(1):17-32. DOI: 10.1007/s10545-009-9006-9

[21] Rehman T, Shabbir MA, Inam-Ur-Raheem M, Manzoor MF, Ahmad N, Liu ZW, et al. Cysteine and homocysteine as biomarker of various diseases. Food Science & Nutrition. 2020;**8**(9):4696-4707. DOI: 10.1002/fsn3.1818

[22] Aguilar B, Rojas JC, Collados MT. Metabolism of homocysteine and its relationship with cardiovascular disease. Journal of Thrombosis and Thrombolysis. 2004;**18**(2):75-87. DOI: 10.1007/s11239-004-0204-x

[23] Selhub J. Homocysteine metabolism. Annual Review of Nutrition. 1999;**19**(1):217-246

[24] Stanger O, Fowler B, Piertzik K, Huemer M, Haschke-Becher E, Semmler A, et al. Homocysteine, folate and vitamin B12 in neuropsychiatric diseases: Review and treatment recommendations. Expert Review of Neurotherapeutics. 2009;**9**(9):1393-1412. DOI: 10.1586/ern.09.75

[25] Karakuła H, Opolska A, Kowal A, Domański M, Płotka A, Perzyński J. Czy dieta ma wpływ na nasz nastrój? Znaczenie kwasu foliowego i homocysteiny [Does diet affect our mood? The significance of folic acid and homocysteine]. Polski Merkuriusz Lekarski. 2009;**26**(152):136-141

[26] Miller JR, Brunold TC. Spectroscopic analysis of the mammalian enzyme cysteine dioxygenase. Methods in Enzymology. 2023;**682**:101-135. DOI: 10.1016/bs.mie.2023.01.002

[27] Lyon P, Strippoli V, Fang B, Cimmino L. B vitamins and one-carbon metabolism: Implications in human health and disease. Nutrients. 2020;**12**(9):2867. DOI: 10.3390/nu12092867

[28] Anam AK, Insogna K. Update on osteoporosis screening and management. The Medical Clinics of North America. 2021;**105**(6):1117-1134. DOI: 10.1016/j.mcna.2021.05.016

[29] Ciancia S, van Rijn RR, Högler W, Appelman-Dijkstra NM, Boot AM, Sas TCJ, et al. Osteoporosis in children and adolescents: When to suspect and how to diagnose it. European Journal of Pediatrics. 2022;**181**(7):2549-2561. DOI: 10.1007/s00431-022-04455-2

[30] Bouxsein ML, Jepsen KJ. Etiology and biomechanics of hip and vertebral fractures. Atlas of Osteoporosis. 2003:165-173

[31] Safadi FF, Barbe MF, Abdelmagid SM, Rico MC, Aswad RA, Litvin J, et al. Bone structure, development and bone biology. Bone Pathology. 2009:1-50

[32] Šromová V, Sobola D, Kaspar P. A brief review of bone cell function and importance. Cell. 2023;**12**(21):2576. DOI: 10.3390/cells12212576

[33] Huiskes R, van Rietbergen BERT. Biomechanics of bone. Basic Orthopaedic Biomechanics and Mechano-biology. 2005;**3**:123-179

[34] Qiu Z-Y, Cui Y, Wang X-M. Natural bone tissue and its biomimetic. In: Mineralized Collagen Bone Graft Substitutes. Sawston, Cambridge: Woodhead Publishing; 2019. pp. 1-22

[35] Pivonka P, Buenzli PR, Dunstan CR. In: Gowder S, editor. A Systems Approach to Understanding Bone Cell Interactions în Health and Disease, Cell Interaction. London: Intech; 2012. DOI: 10.5772/51149. Available from: https://www.intechopen.com/books/cell-interaction/a-systems-approach-to-understanding-bone-cell-interactions-in-health-and-disease

[36] Hadjidakis DJ, Androulakis II. Bone remodeling. Annals of the New York Academy of Sciences. 2006;**1092**:385-396. DOI: 10.1196/annals.1365.035

[37] Gokhale JA, Boskey AL, Robey PG. The biochemistry of bone. Osteoporosis. 2001;**4**:107-188

[38] Kini U, Nandeesh BN. Physiology of bone formation, remodeling, and metabolism. Radionuclide and Hybrid Bone Imaging. 2012:29-57

[39] Wawrzyniak A, Balawender K. Structural and metabolic changes in bone. Animals (Basel). 2022;**12**(15):1946. DOI: 10.3390/ani12151946

[40] Bemben DA, Sherk VD, Ertl WJJ, Bemben MG. Acute bone changes after lower limb amputation resulting from traumatic injury. Osteoporosis International. 2017;**28**(7):2177-2186. DOI: 10.1007/s00198-017-4018-z

[41] Li Z, Kong K, Qi W. Osteoclast and its roles in calcium metabolism and bone development and remodeling. Biochemical and Biophysical Research Communications. 2006;**343**(2):345-350. DOI: 10.1016/j.bbrc.2006.02.147

[42] Szulc P, Bauer DC, Eastell R. Biochemical markers of bone turnover in osteoporosis. In: Marcus and Feldman's Osteoporosis. Cambridge, Massachusetts: Academic Press; 2021. pp. 1545-1588

[43] McClung MR. The relationship between bone mineral density and fracture risk. Current Osteoporosis Reports. 2005;**3**(2):57-63

[44] Bessette L et al. Factors influencing the treatment of osteoporosis following fragility fracture. Osteoporosis International. 2009;**20**:1911-1919

[45] World Health Organization. Prevention and Management of Osteoporosis: Report of a WHO Scientific Group. No. 921. Geneva: World Health Organization; 2003

[46] Ambrose AF, Paul G, Hausdorff JM. Risk factors for falls among older adults: A review of the literature. Maturitas. 2013;**75**(1):51-61

[47] Geetha J. Prevalence of falls and associated factors in elderly population [diss]. Chennai: Madras Medical College; 2015

[48] Marks R. Falls among the elderly and vision: A narrative review. Open Medicine Journal. 2014;**1**:1

[49] Bliuc D, Nguyen ND, Nguyen TV, Eisman JA, Center JR. Compound risk of high mortality following osteoporotic fracture and re-fracture în elderly women and men. Journal of Bone and Mineral Research. 2013;**28**(11):2317-2324

[50] Pisani P, Renna MD, Conversano F, Casciaro E, Di Paola M, Quarta E, et al. Major osteoporotic fragility fractures: Risk factor updates and societal impact. World Journal of Orthopaedics. 2016;**7**(3):171-181

[51] Adams AL, Adams JL, Raebel MA, Tang BT, Kuntz JL, Vijayadeva V, et al. Bisphosphonate drug holiday and fracture risk. Journal of Patient-Centered Research and Reviews. 2015;**2**:101

[52] Reid IR, Bolland MJ. Calcium and/or vitamin D supplementation for the prevention of fragility fractures: Who needs it? Nutrients. 2020;**12**(4):1011

[53] Tran T, Bliuc D, Pham HM, van Geel T, Adachi JD, Berger C, et al. A risk assessment tool for predicting fragility fractures and mortality în the elderly. Journal of Bone and Mineral Research. 2020;**35**(10):1923-1934

[54] Sellmeyer D, Civitelli R, Hofbauer LC, Khosla S, Lecka-Czernik B, Schwartz AV. Skeletal metabolism, fracture risk, and fracture outcomes in type 1 and type 2 diabetes. Diabetes. 2016;**65**(7):1757-1766. DOI: 10.2337/db16-0063

[55] Dargent-Molina P et al. Fall-related factors and risk of hip fracture: The EPIDOS prospective study. The Lancet. 1996;**348**(9021):145-149

[56] Ross PD. Risk factors for osteoporotic fracture. Endocrinology and Metabolism Clinics of North America. 1998;**27**(2):289-301

[57] Van der Voort DJM, Geusens PP, Dinant GJ. Risk factors for osteoporosis related to their outcome: Fractures. Osteoporosis International. 2001;**12**:630-638

[58] Richards JB et al. Bone mineral density, osteoporosis, and osteoporotic fractures: A genome-wide association study. The Lancet. 2008;**371**(9623):1505-1512

[59] Palermo A et al. BMI and BMD: The potential interplay between obesity and bone fragility. International Journal of Environmental Research and Public Health. 2016;**13**(6):544

[60] Rinonapoli G et al. Obesity and bone: A complex relationship. International Journal of Molecular Sciences. 2021;**22**(24):13662

[61] Miyao M, Morita H, Hosoi T, Kurihara H, Inoue S, Hoshino S, et al. Association of methylenetetrahydrofolate reductase (MTHFR) polymorphism with bone mineral density in postmenopausal Japanese women. Calcified Tissue International. 2000;**66**(3):190-194. DOI: 10.1007/s002230010038

[62] van Meurs JB, Dhonukshe-Rutten RA, Pluijm SM, van der Klift M, de Jonge R, Lindemans J, et al.

Homocysteine levels and the risk of osteoporotic fracture. The New England Journal of Medicine. 2004;**350**(20):2033-2041. DOI: 10.1056/NEJMoa032546

[63] Gjesdal CG, Vollset SE, Ueland PM, Refsum H, Drevon CA, Gjessing HK, et al. Plasma total homocysteine level and bone mineral density: The Hordaland homocysteine study. Archives of Internal Medicine. 2006;**166**(1):88-94. DOI: 10.1001/archinte.166.1.88

[64] Périer MA, Gineyts E, Munoz F, Sornay-Rendu E, Delmas PD. Homocysteine and fracture risk in postmenopausal women: The OFELY study. Osteoporosis International. 2007;**18**(10):1329-1336. DOI: 10.1007/s00198-007-0393-1. Epub 2007 Jun 5

[65] Rumbak I, Zižić V, Sokolić L, Cvijetić S, Kajfež R, Colić BI. Bone mineral density is not associated with homocysteine level, folate and vitamin B12 status. Archives of Gynecology and Obstetrics. 2012;**285**(4):991-1000. DOI: 10.1007/s00404-011-2079-3

[66] Jianbo L, Zhang H, Yan L, Xie M, Mei Y, Jiawei C. Homocysteine, an additional factor, is linked to osteoporosis in postmenopausal women with type 2 diabetes. Journal of Bone and Mineral Metabolism. 2014;**32**(6):718-724. DOI: 10.1007/s00774-013-0548-4

[67] Sandeep MMR, Prabharam KR. A prospective study to access the co-relation between osteoporosis and serum homocysteine level. International Journal of Orthopaedics Sciences. 2020;**6**(4):93-96. DOI: 10.22271/ortho.2020.v6.i4b.2325

[68] Garcia-Alfaro P, Rodriguez I, Pascual MA. Evaluation of the relationship between homocysteine levels and bone mineral density in postmenopausal women. Climacteric. 2022;**25**(2):179-185. DOI: 10.1080/13697137.2021.1921729

[69] Widjaja SS, Rusdiana R, Jayalie VF. Homocysteine levels and osteoporotic fracture in a population aged 55 years over in Medan District Indonesia. Open access Macedonian. Journal of Medical Sciences. 2022;**10**(A):45-48

[70] Watanabe N, Ogawa T, Takada R, Amano Y, Jinno T, Koga H, et al. Association of osteoporosis and high serum homocysteine levels with intraoperative periprosthetic fracture during total hip arthroplasty: A propensity-score matching analysis. Archives of Orthopaedic and Trauma Surgery. 2023;**143**(12):7219-7227. DOI: 10.1007/s00402-023-04989-6

[71] Enneman AW, Swart KM, Zillikens MC, van Dijk SC, van Wijngaarden JP, Brouwer-Brolsma EM, et al. The association between plasma homocysteine levels and bone quality and bone mineral density parameters in older persons. Bone. 2014;**63**:141-146. DOI: 10.1016/j.bone.2014.03.002

[72] Wang C, Zhang X, Qiu B. Genetically predicted circulating serum homocysteine levels on osteoporosis: A two-sample mendelian randomization study. Scientific Reports. 2023;**13**(1):9063. DOI: 10.1038/s41598-023-35472-2

[73] Koh JM, Lee YS, Kim YS, Kim DJ, Kim HH, Park JY, et al. Homocysteine enhances bone resorption by stimulation of osteoclast formation and activity through increased intracellular ROS generation. Journal of Bone and Mineral Research. 2006;**21**(7):1003-1011. DOI: 10.1359/jbmr.060406

[74] Miyamoto T, Hirayama A, Sato Y, Koboyashi T, Katsuyama E, Kanagawa H, et al. A serum metabolomics-based

profile in low bone mineral density postmenopausal women. Bone. 2017;**95**:1-4. DOI: 10.1016/j.bone.2016.10.027

[75] Zhao Q, Shen H, Su KJ, Zhang JG, Tian Q, Zhao LJ, et al. Metabolomic profiles associated with bone mineral density in US Caucasian women. Nutrition & Metabolism (London). 2018;**15**:57. DOI: 10.1186/s12986-018-0296-5

[76] Yamada M, Tsukimura N, Ikeda T, Sugita Y, Att W, Kojima N, et al. N-acetyl cysteine as an osteogenesis-enhancing molecule for bone regeneration. Biomaterials. 2013;**34**(26):6147-6156. DOI: 10.1016/j.biomaterials.2013.04.064

[77] Wang Y, Han X, Shi J, Liao Z, Zhang Y, Li Y, et al. Distinct metabolites in osteopenia and osteoporosis: A systematic review and meta-analysis. Nutrients. 2023;**15**(23):4895. DOI: 10.3390/nu15234895

Chapter 4

Deep Vein Thrombosis in Pregnancy and Postpartum: Are Sulfur-Containing Amino Acids Involved in Thrombophilia Condition?

Cristiana Filip, Catalina Filip, Roxana Covali, Mihaela Pertea, Daniela Matasariu, Gales Cristina and Demetra Gabriela Socolov

Abstract

Thrombophilia is a life-threatening condition causing deep vein thrombosis associated with pulmonary thromboembolism. In pregnancy and postpartum, the risk of venous thromboembolism is 5 times higher; in association with pre-existing thrombophilia becoming up to 30 times higher. The main cause of mortality at birth in underdeveloped countries is hemorrhage, while in developed countries, mortality is caused by thromboembolic complications. A peculiarity of pregnancy nowadays is the advanced age of the mother at the time of conception and assisted reproduction, both conditions presenting thrombotic risks through hyperstimulation that favors hemoconcentration as a result of high levels of estradiol generation and/or immobilization, which favors hypercoagulability and DVT respectively. In this chapter, we have summarized the most important connection between thrombophilia, deep vein thrombosis and Hcy involvement in pregnancy and postpartum conditions.

Keywords: cysteine, homocysteine, thrombophilia, pregnancy, deep vein thrombosis

1. Introduction

Pregnancy is characterized by physiologic changes that lead to a relative hypercoagulable state, which is accompanied by an increased stasis. The procoagulant status can be significantly amplified in the case of inherited or acquired thrombophilia, autoimmune diseases, or particular medical conditions such as prolonged immobilization. The major risk is encountered, in most cases, in the postpartum period when the incidence of venous thromboembolism increases up to 2.5 times [1]. Thus, the literature indicates a number of 500 per 100,000 venous thromboembolic events in postpartum compared to 200 per 100,000 during pregnancy [2]. In developed countries, mortality at birth is caused by thromboembolic complications compared to underdeveloped countries where the mortality is caused by hemorrhage [3]. Thus, venous thromboembolic events are very serious medical issues which have been estimated to range from 1.2 to 4.7 per 100,000 pregnancies [1]. The risk of venous thromboembolism in pregnancy is up to 30 times higher when it is associated with acquired or hereditary thrombophilia [4, 5]. During pregnancy, up to 50% of patients presenting a thrombotic event were found to have thrombophilia [1]. Currently, a new risk of venous thromboembolism has appeared for two reasons: firstly, the age of first pregnancy has moved toward 35–45 years and secondly, the number of assisted pregnancies has increased significantly. In assisted pregnancies, thrombotic risk increases due to fertilization techniques that include hormonal hyperstimulation, a procedure that promotes hypercoagulability and deep vein thrombosis (DVT). In other words, due to this combination of factors, the risk of venous thrombosis is greatly increased these days. One of the thrombogenic factors that stands out is the methylene-tetrahydrofolate reductase (MTHFR) enzyme mutation. This mutation affects the metabolism of amino acids containing the sulfur atom, namely the metabolism of methionine. As a consequence of the disruption of methionine metabolism, homocysteine (Hcy) accumulates. Homocysteine is a non-proteinogenic amino acid involved in the pathogenesis of vascular endothelium and promoter of thrombogenicity. It also represents the main precursor in Cysteine (Cys) synthesis the only proteinogenic amino acid containing a free thiol group. The structural similarity between homocysteine and cysteine makes the latter interesting to investigate for possible involvement in thrombophilia and deep vein thrombosis.

2. Coagulation parameters during pregnancy

Physiological changes that occur during pregnancy may favor a certain level of thrombogenicity without necessarily major events (**Figure 1**). In contrast, postpartum thrombogenicity, which aims to avoid hemorrhage after delivery, can cause severe thrombotic events. The coagulation parameters that undergo a series of physiological changes are:

- activated clotting factors: high levels for factors VII, VIII, X, von Willebrand factor, and fibrinogen
- decreased anticoagulant factors: protein S level [6], adapted resistance to activated protein C
- inhibiting fibrinolytic factors: increased activatable fibrinolytic inhibitor (TAFI), plasminogen activator inhibitor 1 and 2 (PAI-1 and PAI-2).

When the particular condition of pregnancy overlaps with inherited or acquired thrombophilia, it is very likely that the thrombogenic effect will be significantly amplified. Inherited thrombophilia is caused either by deficiencies or by mutations of specific enzymes in the coagulation pathway. Because clotting factors are plasma-functioning enzymes, also known as zymogens, their dysfunction leads to clotting defects. Currently, three important inherited thrombophilia are considered responsible for the majority of thromboembolic events [7], namely:

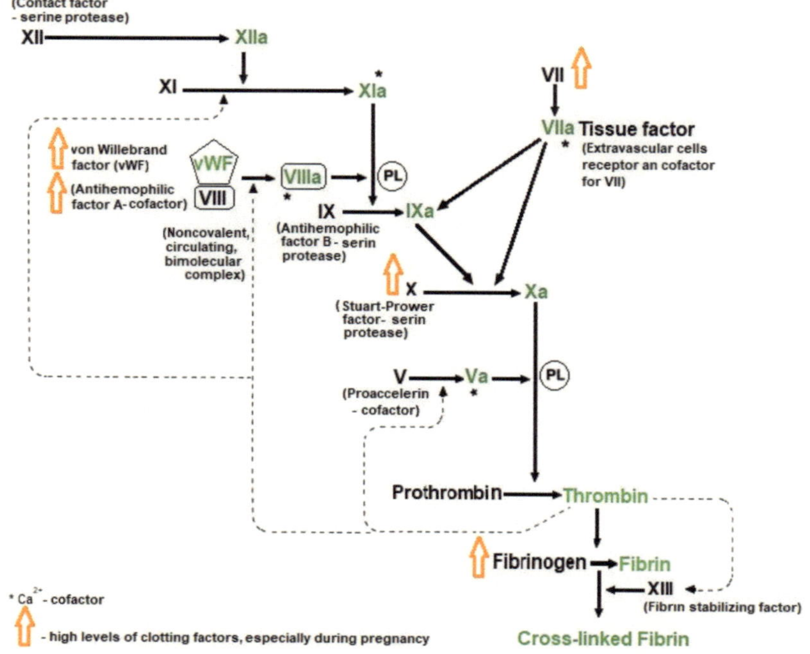

Figure 1.
Physiological changes in the coagulation parameters during pregnancy.

a. a mutation of the factor V gene (factor V Leiden), causes resistance to activated protein C (**Figure 2**). The mutation is considered the major cause of venous thrombosis and responsible for 20–30% of venous thromboembolism events [8].

b. mutation of the prothrombin gene leads to higher concentrations of plasma prothrombin and increased risk of venous thromboembolism (**Figure 2**).

c. mutation of the methylenetetrahydrofolate reductase (MTHFR) gene results in increased homocysteine plasma concentrations. The mutation causes low MTHFR activity and can be found in 5–15% of the population.

a. Protein C (PC) is an anticoagulant serine protease synthesized by the liver that circulates in plasma as an inactive precursor. In its activated form, protein C (aPC) proteolytically inhibits the activities of coagulation factors Va and VIIIa (**Figure 2**). The activation of PC is triggered by the thrombin-thrombomodulin complex, located on the luminal surface of endothelial cells of blood vessels. Protein C binding to the endothelial cell Protein C receptor (EPCR) increases the efficiency of the process. Protein C activity also requires protein S as a cofactor. Once activated, protein C proteolytically degrades activated coagulation factors Va and VIIIa. Thus, within the coagulation process, activated protein C down-regulates the coagulation cascade through limited proteolytic degradation of cofactors Va and VIIIa. As a result, the amplification and progression of the coagulation cascade decrease and capillary permeability is maintained [9].

Factor V contains three cleavage sites on which activated protein C acts. Each of these three sites contains Arg in the following positions Arg306, Arg506, and

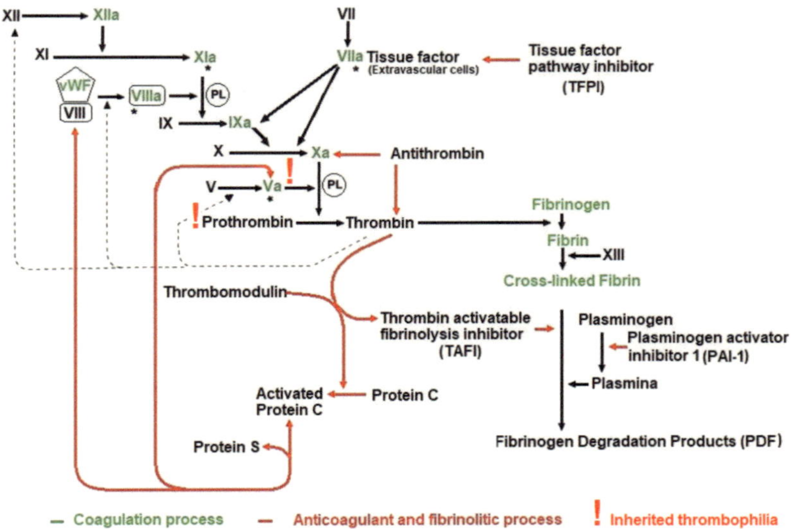

Figure 2.
Inherited thrombophilia responsible for thrombogenesis.

Arg 679 respectively. Following the action of aPC, factor V is proteolytically degraded, thus blocking the coagulation cascade. Dysfunctional factor V (called factor V Leiden) is caused by a point mutation in the coagulation factor V gene that substitutes Arg 506 for Gln. The presence of Glu prevents the proteolytic action of aPC, thus allowing coagulation to continue [10–12]. The literature [13] states that approximately 90% of cases with activated protein C resistance are caused by factor V gene mutation.

b. Thrombin is a turntable in the process of coagulation, on one hand, it activates clot formation by transforming fibrinogen into fibrin, on the other hand, together with thrombomodulin, it controls the extent and intensity of coagulation.

Thrombin is generated from prothrombin (PTM), and the prothrombin G20210A gene mutation is the second most commonly inherited thrombophilia after Factor V Leiden. The prothrombin mutation is strongly accompanied by increased levels of serum prothrombin and thromboembolic events [14, 15].

Prothrombin is activated to thrombin mainly by factor Xa. The resulting α-thrombin molecule is composed of two chains, A and B respectively, held together by a disulfide bond formed between two cysteine residues (Cys293-Cys439). Mutations identified in this domain lead to prothrombin dysfunction [16]. Thrombin activity is controlled by antithrombin (AT), a serine protease, that inactivates thrombin. Antithrombin is synthesized by the liver, contains three disulfide bonds in its structure and a deficiency of it is associated with venous thrombosis. The literature shows that proper folding of antithrombin and the formation of disulfide bonds between the appropriate cysteine amino acids is essential for its activity [17]. The AT function is conditioned by two important

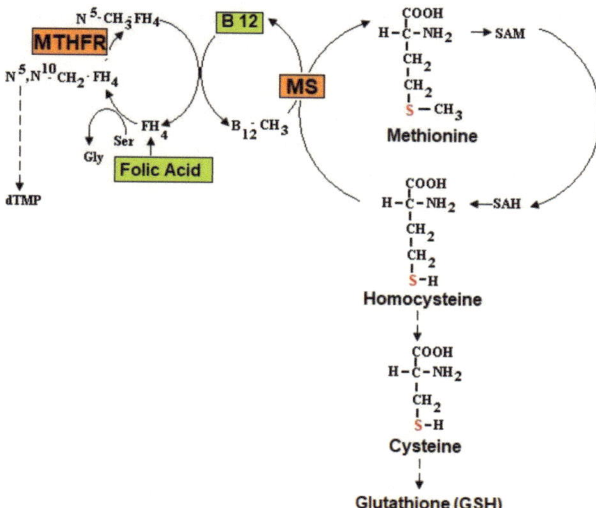

Figure 3.
Short metabolism of sulfur-containing amino acids.

structural features, namely the presence of a central loop that allows covalent binding to the active site of coagulation proteases thus inactivating them, and the presence of a structural motif (basic D-helix) that allows binding to heparan sulfate proteoglycans (HSPG) on the vascular wall [18].

c. Mutation of the methylenetetrahydrofolate reductase (MTHFR) gene (**Figure 3**) results in increased homocysteine plasma concentrations largely accepted as a risk factor for both arterial occlusive disease and venous thrombosis. Homocysteine is a sulfur-containing amino acid and the higher homolog of cysteine. Unlike cysteine, which is a proteinogenic amino acid, homocysteine is a non-proteinogenic amino acid but homocysteine metabolism is a major pathway for cysteine synthesis.

From an epidemiologic point of view, polymorphisms of the methylene tetrahydrofolate reductase (MTHFR) gene are common, literature indicates up to 40% of the general population [19]. MTHFR gene polymorphism is the main cause of hyperhomocysteinemia (HHcy), up to 20% are homozygous for MTHFR 677C > T or 1298A > C.

The risk of thrombosis and MTHFR variants is still debatable as homocysteine levels are impacted by many factors, including concomitant renal disease, thyroid disease, nutritional deficiencies, and alcohol intake.

3. Cysteine involvement in pregnancy

Cysteine is a unique human proteinogenic amino acid that contains a reactive free sulfhydryl (–SH) or thiol group. Cysteine is crucial for protein synthesis (generates disulfide bonds in tertiary and quaternary structures), for redox homeostasis (provides the active component (-SH) in the major antioxidant agent in humans, namely glutathione), or for other metabolic functions (main component in acetyl-coenzyme A, a vital structure in energy generation). Very high doses (more than 7 grams) of cysteine may be toxic to human cells and may lead to death [20, 21]. Unlike homocysteine, which is associated with thrombogenicity, very little is known about cysteine regarding its involvement in both general and coagulation pathology. However, it is well known that a deficiency in the reabsorption of cystine (two Cys molecules linked together by a disulfide bridge) in the renal tubules can lead to stone formation and early renal failure.

The literature does not mention a direct involvement of cysteine in pregnancy or postpartum but emphasizes the effect of N-acetylcysteine (NAC) administration, in different periods of pregnancy or in the pathology encountered during pregnancy. N-acetylcysteine is a cysteine derivative in which the alpha-amino group is acetylated and the thiol group stays free. This allows the thiol group to intervene in the detoxification process as well as other processes in the body. Amin and coworkers [22] mentioned the benefit of NAC administration in recurrent pregnancy loss. The authors consider oxidative stress as the main cause that triggers a cascade of changes that may lead to pregnancy loss. As the thiol group of cysteine within NAC remains free, it is able to prevent oxidative stress by decreasing oxidative genotoxicity and by inhibiting the release of proinflammatory cytokines. Authors also find that NAC seems to mitigate the placental apoptosis and inflammatory responses that are related to severe oxidative stress. As a conclusion, authors mentioned that the use of N-acetyl cysteine in therapy enhances pregnancy outcomes in patients diagnosed with recurrent unexplained pregnancy loss (RPL) [22].

Other researchers consider that oxidative stress may be involved in the incidence of preeclampsia (PE), one of the most common pregnancy disorders [23]. N-acetyl cysteine is believed to have strong antioxidant properties and has been reported to be beneficial in preventing and treating oxidative stress in preeclampsia [24]. As a conclusion, N-acetyl cysteine supplementation is associated with a decrease in elevated blood pressure and proteinuria, which are hallmarks of preeclampsia [25].

Long-term unexplained infertility is another pathology mentioned in the literature where administration of N-acetyl cysteine appears to be beneficial. Researchers hypothesized that the effects of N-acetyl-cysteine such as: insulin-sensitizing properties, antiapoptotic and antioxidant effects, and protection against focal ischemia, may have ovulation-enhancing effects. Thus, researchers conclude that NAC is an effective, cheap, and safe adjuvant that improves pregnancy rate significantly [26, 27].

However latest data present information about oxidizing processes undertaken by cysteine and methionine (**Figure 4**) with thrombogenic consequences [28]. These data suggest that cysteine and methionine oxidation (**Figure 4**) may act as

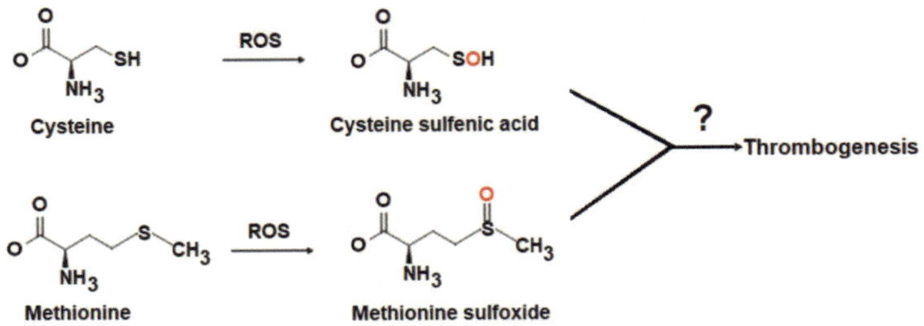

Figure 4.
Cysteine and methionine forms oxidized under reactive oxygen species (ROS) supposed to be involved in thrombogenesis.

prothrombotic regulators. It seems that proteins involved in coagulation/thrombotic processes such as fibrinogen, von Willebrand factor, or β2 glycoprotein I may undertake post-translational oxidation thus promoting thrombogenesis [29–33].

Moreover, a very recent study emphasizes the importance of sulfur-containing amino acids, especially cysteine. During pregnancy, free cysteine thiol groups belonging to certain proteins are converted to persulfides to generate hydrogen sulfide (H_2S). Although the exact role of H_2S is not yet known, it is hypothesized to be involved in stimulating vital systemic/local vasodilator processes that contribute to uterine vasodilation during pregnancy and to the regulation of vascular smooth muscle contraction/relaxation. Thus, H_2S is involved in the sulfhydration of specific proteins that are responsible for regulating uterine hemodynamics [34].

To resume, the oxidation of sulfur-containing amino acids is considered a possible link between oxidative processes and thrombogenesis. On the other hand, a new role of the -SH group of cysteine in uterine hemodynamics has recently been reported.

As a conclusion, it emerges a wider involvement of cysteine in other cellular processes besides oxidative ones, which opens "new insights" in investigating the complex role of cysteine.

4. Homocysteine and pregnancy complication

Homocysteine is an amino acid generated by the human body in the methionine metabolism. It does not participate in protein synthesis, but it is an important intermediate metabolite in major processes such as transmethylation, trans-sulfuration, cysteine formation, etc. In the transmethylation process homocysteine participates in the formation of adrenaline, melatonin, lecithin, creatine, etc. The trans-sulfuration pathway leads to cysteine synthesis, amino acid that plays a key role in the spatial conformations of protein and particularly in glutathione generation, the most important antioxidant agent in the body (**Figure 3**).

Homocysteine metabolism presented in **Figure 3** highlights the involvement of two vitamins (folic acid and vitamin B_{12}) and two enzymes methylene-tetrahydrofolate reductase (MTHFR) and methionine synthase (MS) respectively. The absence of these vitamins as well as the enzyme deficiencies trigger the increase of Hcy levels. The normal concentration of homocysteine in human blood is 5–15 μM. Over the past 40 years, homocysteine has been noticed in various pathologies where its concentrations exceed the above range. This condition is known as hyperhomocysteinemia (HHcy). High homocysteine concentrations are classified according to clinical consequences as being moderate at 16–30 μM, intermediary at 31–100 μM, and severe above 100 μM [35]. Nowadays it is generally accepted that HHcy promotes thrombosis [36–38].

Nowadays there are two opposite opinions regarding the role of homocysteine in pathology: first theory considers homocysteine as a risk factor [39] while the second considers Hcy as being the marker of chronic vascular injury [40]. Both points of view bring scientific arguments and, against the fact that the controversy continues, it is generally accepted that homocysteine is involved in vascular endothelial dysfunction and promotes thrombosis [41–43]. The literature mentions many diseases in which elevated homocysteine concentrations have been found in the blood: cardiovascular diseases [36, 44–48], neurological diseases [49–52], miscarriage [53–55], thrombophilia [56–61], bone fragility [62–67], diabetes [68–74], inflammation [75–77].

The link between homocysteinemia and pregnancy complications is mediated by inherited/acquired MTHFR enzyme mutations (**Figure 3**). This mutation results in hyperhomocysteinemia, which causes thrombophilia. Miscarriage is often encountered in this condition. Clinical studies report that this mutation causes spontaneous abortions 3.3 times more frequently [53, 54]. The particular case of the simultaneous mutation of MTHFR C667T, factor V Leiden, and/or prothrombin gene mutations will definitely cause recurrent pregnancy loss [55]. In addition to the thrombogenic effect, HHcy also manifests an effect of diminishing the fibrinolytic process. Fibrinolysis is controlled by different proteins; among them, protein C plays a key role in this process. Protein C has anticoagulant activity by blocking the activity of factors V and VIII. Thus, a low level of protein C will favor the formation of clots. Data from the literature indicate that protein C activity is affected by increased levels of Hcy as a result of the formation of disulfide bridges between the two [56]. Consequently, a thrombogenic status is promoted which will favor the formation of venous and arterial clots [57–60]. Therefore, a high level of homocysteine can lead to the formation of clots, deep vein thrombosis, or acute events such as pulmonary thromboembolism and also to pregnancy loss. These mechanisms are extensively presented in our previously published work [61].

5. Homocysteine mechanism of action

The mechanisms by which hyperhomocysteinemia triggers injury processes are thought to be as follows: oxidative stress [78–84], inflammation [85–95], and cellular signaling [96–113]. High concentrations of homocysteine in the blood are considered to be harmful to vascular endothelial function.

In a physiological state, the endothelium has anticoagulant and vasorelaxant activity that favors exchanges at the cellular level. Endothelial function impairment triggers both functional (coagulability, proinflammatory cytokine secretion) and morphological (vascular hypertrophy) imbalance. As a result of injury, whatever the cause, the vascular endothelium responds through a chain of events leading to oxidative stress, intracellular signaling, and cellular renewal. The exact point or mechanism by which HHcy intercepts this chain of events has not been clearly identified. However, the generation of reactive oxygen species [75], activation of the inflammatory response [76], and cell signaling [77] are considered possible mechanisms and are widely investigated at present. Some of these scientific hypotheses will be briefly described below.

5.1 Hyperhomocysteinemia involvement in oxidative stress

Scientific evidence suggests a significant role of hyperhomocysteinemia in disrupting cellular redox balance in the vascular endothelium. The specific reactions that mediate the vascular effects of HHcy involve the synthesis of reactive oxygen species (ROS). Some of these species are themselves intracellular messengers (such as hydrogen peroxide) [78–80]. The literature highlights the fact that HHcy generates reactive species both by autooxidation and by direct reaction with other cell structures or components [81, 82]. Our studies in an experimentally induced HHcy model in rats showed that HHcy promotes the generation of hydrogen peroxide and decreases the total antioxidant capacity of serum [83, 84].

The oxidative process leads to the injury of endothelial cells [114], diminished blood vessels in villi, and decreases in pregnancy, the circulation of blood at the maternal-fetal interface. Additionally, HHcy reduces NO releasing by the endothelial cells, which affect placental perfusion [115], and induces platelet accumulation, and promotes thrombosis [116].

5.2 Hyperhomocysteinemia involvement in inflammation

Cell survival depends on responding to, eliminating, and then returning to the original, healthy structure after any type of insult or injury. To do this, cells have developed over time a complex system called the inflammasome. The inflammasome is an integrative structure that collects data about aggression, mobilizes and coordinates the response accordingly, and finally restores the initial morphological and physiological status. In addition, the inflammasome stores information about aggression to be quickly accessed in case of a future attack.

The coordination of the whole process is achieved by the activation/secretion of specific messengers of the immune system, namely cytokines [85]. Once secreted, these molecules trigger the formation of reactive species, the modification of ion fluxes, which will ultimately lead to the modulation of gene transcription in order to supply the components/proteins required for the entire process.

Recent data identify HHcy as the key factor in the formation and activation of the Nod-like receptor protein 3 complex (NLRP3) of the inflammasome [86, 87].

Current data report [88, 89] HHcy as promoting chronic inflammation in endothelial cells. Chronic inflammation is an irreversible injury, in other words, it is an incurable injury, due to its particular characteristics. This means that chronic inflammation induces abnormal morphological changes not only in the area of injury but also in the surrounding tissues, ultimately altering their normal physiology. Constantly increased levels of proinflammatory parameters such as interleukins (IL-6, IL-8, and TNFα) [90, 91] have been identified in chronic inflammation and they are positively associated with hyperhomocysteinemia. Elevated homocysteine levels stimulate IL-1beta and TNF-alpha secretion in human monocytes/macrophages. Overall hyperhomocysteinemia seems to alter the profile of cytokines produced by both endothelial cells and macrophages [92–95].

5.3 Hyperhomocysteinemia involvement in cellular signaling

A causal loop is formed between inflammation and ROS, as they stimulate each other. In other words, inflammation triggers the synthesis of reactive species that act as messengers/mediators and trigger the immune system cell response. Thus, the immune cells will produce specific molecules that will act both on themselves and on other target cells. After activation and self-activation, the target cells will produce reactive species in turn. Cellular survival depends on this functional loop. In addition, inflammasome-reactive species connection functions in both pathological and physiological processes [96].

The levels of ROS concentrations make the difference between the physiological regulatory pathways and the pathological process called oxidative stress; the latter being associated with high ROS concentration [97]. Reactive species can be signals/stimuli in pathological processes, but cells can also produce them to use in intercellular signaling processes. The reactive species group includes oxygen superoxide, the hydroxyl radical, and hydrogen peroxide (H_2O_2). Currently, H_2O_2 is considered a second messenger capable of crossing the cell membrane due to its partial lipophilic character [98]. Scientific research has shown that hydrogen peroxide has similar activity to growth factors when is added to a living system. In turn, growth factors can trigger the release of hydrogen peroxide at the cellular level [99–101]. According to scientific data, ROS including H_2O_2 are involved in the MAPK signaling pathway that ultimately acts on gene transcription [102–107]. HHcy generates hydrogen peroxide and, as a consequence, may indirectly intervene in the signaling process. Thus, hyperhomocysteinemia is a potential activator of the mitogenic process [108–113].

The literature also mentions another effect of HHcy, decreased NO release that disrupts vascular relaxation and alters the anticoagulant activity of protein C [117]. Therefore, HHcy promotes endothelial injury leads to inflammation and oxidative process and finally to thrombosis [118].

As a conclusion, in a high homocysteine status, the vascular function is disturbed thus generating an increased risk of thrombosis [57–60]. As a consequence, HHcy is associated with pregnancy complications.

6. Homocysteine blood level in pregnancy

The blood Hcy concentration during normal pregnancy varies depending on the race population as seen in **Table 1**. M. Walker [119, 120]:

Literature mentions that pregnant women with inherited thrombophilia are more susceptible to HHCy [121, 122]. Hyperhomocysteinemia is not very common but is an important cause of deep vein thrombosis and recurrent pregnancy loss. Serum HHcy is associated with preeclampsia, recurrent miscarriage, and low birth weight for the newborn. Even so, the investigations of serum homocysteine levels are not of routine investigations, and diagnosis of this condition is missed due to an extremely rare evaluation of serum homocysteine levels. The literature mentions a case report in which an abnormally high plasma homocysteine level (26.58 μmol/L) was found following a spontaneous abortion at 28 weeks gestation. The patient had a history of two unexplained pregnancy losses without having previously been investigated for Hcy levels [123].

Langman et al. also mentioned in a retrospective case-control study that hyperhomocysteinemia is a major risk factor for venous thromboembolic disease and recurrent abortion [124].

The postpartum period is conventionally defined as 6 weeks after birth. This period is considered to have the highest risk of thrombosis. The literature mentions that the risk of postpartum thrombosis can persist up to 12 weeks after delivery [125].

However, regarding Hcy levels in the postpartum period, there are almost no data in the literature. Even so, the determination of serum Hcy levels during pregnancy is

Period of pregnancy (weeks)	Hcy blood concentrations (according to Walker) (mmol/L)	Hcy blood concentrations (according to Y. Yang) (μmol/L)
up to 16	3.9–7.3	5.79–11.86
20–30	3.5–5.3	5.79–11.86
after 36	3.3–7.5	6.13–16.75

Table 1.
Blood Hcy concentration during normal pregnancy.

recommended because it can indicate the mother's metabolic status and is useful in predicting the risk of adverse pregnancy and postpartum events.

Hcy is an early warning for most complications encountered during pregnancy (miscarriage, preeclampsia), but can be a warning sign for postpartum as well.

7. Conclusion

Among sulfur-containing amino acids, homocysteine stands out because of its association with high levels of thrombosis. This association can be explained by a multifactorial mechanism, which ultimately disturbs the balance between pro-coagulation and anticoagulation factors. Coexistence of a deficiency in Hcy metabolism, such as methylenetetrahydrofolate reductase together with the mutation of some coagulation factors such as factor V Leiden and prothrombin, leads to thrombogenic risk and venous thromboembolism as well as pregnancy complications. Therefore, in addition to the investigation of coagulation parameters, we consider that blood homocysteine levels should be introduced as a routine investigation in the diagnosis of pregnant women with pregnancy complications and deep vein thrombosis.

Conflict of interest

The authors declare no conflict of interest.

Author details

Cristiana Filip[1*], Catalina Filip[2], Roxana Covali[3], Mihaela Pertea[4], Daniela Matasariu[5], Gales Cristina[6] and Demetra Gabriela Socolov[7]

1 Department of Morpho-Functional Sciences (II), Discipline of Biochemistry, University of Medicine and Pharmacy "Grigore T. Popa", Iasi, Romania

2 Chirurgie vasculaire et endo-vasculaire, CHU Gabriel-Montepied, Clermont-Ferrand, France

3 Department of Fundamental Biomedical Sciences, University of Medicine and Pharmacy "Grigore T. Popa", Iasi, Romania

4 Department of Plastic Surgery and Reconstructive Microsurgery, University of Medicine and Pharmacy "Grigore T. Popa", Iasi, Romania

5 Department of Maternal and Child Medicine, University of Medicine and Pharmacy "Grigore T. Popa", Iasi, Romania

6 Department of Morpho-Functional Sciences (I), Discipline of Histology, University of Medicine and Pharmacy "Grigore T. Popa", Iasi, Romania

7 Department of Obstetrics and Gynecology, University of Medicine and Pharmacy "Grigore T. Popa", Iasi, Romania

*Address all correspondence to: cristiana.filip@umfiasi.ro

References

[1] Battinelli EM, Marshall A, Connors JM. The role of thrombophilia in pregnancy. Thrombosis. Dec 2013;**2013**:516420

[2] Heit JA, Kobbervig E, James AH, Petterson TM, Bailey KR, Melton LJ III. Trends in the incidence of venous thromboembolism during pregnancy or post-partum: A 30-year population-based study. Annals of Internal Medicine. 2005;**143**(10):697-706

[3] Devis P, Knuttinen MG. Deep venous thrombosis in pregnancy: Incidence, pathogenesis and endovascular management. Cardiovascular Diagnosis and Therapy. 2017;**7**(Suppl. S3):S309-S319

[4] De Jong PG, Coppens M, Middeldorp S. Duration of anticoagulant therapy for venous thromboembolism: Balancing benefits and harms on the long term. British Journal of Haematology. 2012;**158**:433-441

[5] Marik PE, Plante LA. Venous thromboembolic disease and pregnancy. The New England Journal of Medicine. 2008;**359**:2025-2033

[6] Bremme A. Haemostatic changes in pregnancy. Best Practice & Research. Clinical Haematology. 2003;**16**(2):153-168

[7] Kupferminc MJ. Thrombophilia and pregnancy. Reproductive Biology and Endocrinology. 2003;**1**:111

[8] Bertina RM, Koeleman RPC, Koster T, Rosendaal FR, Dirven RJ, de Ronde H, et al. Mutation in blood coagulation factor V associated with resistance to activated protein C. Nature. 1994;**369**(6475):64-67

[9] Stojanovski BM, Pelc LA, Di Cera E. Role of the activation peptide in the mechanism of protein C activation. Scientific Reports. 2020;**10**(1):11079

[10] Green D, Maliekel K, Sushko E, Akhtar R, Soff GA. Activated-protein-C resistance in Cancer patients. Pathophysiology of Haemostasis and Thrombosis. 1997;**27**(3):112-118

[11] Castoldi E, Rosing J. APC resistance: Biological basis and acquired influences. Journal of Thrombosis and Haemostasis. 2010;**8**:445-453

[12] Lisa FL, Lonergan A, Scorgie FE, Rowlings P, Gibson R, Lawrie A, et al. Endogenous thrombin potential for predicting risk of venous thromboembolism in carriers of factor V Leiden. Pathophysiology of Haemostasis and Thrombosis. 2006;**35**(6):435-439

[13] Moore GW, Castoldi E, Teruya J, Morishita E, Adcock DM. Factor V Leiden-independent activated proteinC resistance: Communication from the plasma coagulation inhibitors subcommittee of the international society on thrombosis and haemostasis scientific and standardisation committee. Journal of Thrombosis and Haemostasis. 2023;**21**:164-174

[14] Elkattawy S, Alyacoub R, Singh KS, Fichadiya H, Kessler W. Prothrombin G20210A gene mutation-induced recurrent deep vein thrombosis and pulmonary embolism: Case report and literature review. Journal of Investigative Medicine High Impact Case Reports. 2022;**10**:23247096211058486

[15] Jayandharan G, Viswabandya A, Baidya S, Nair SC, Shaji RV, Chandy M, et al. Molecular genetics of hereditary prothrombin deficiency in Indian

patients: Identification of a novel Ala362 -> Thr (prothrombin Vellore 1) mutation. Journal of Thrombosis and Haemostasis. 2005;**3**(7):1446-1453

[16] Beretta AL, Bianchi M, Norchi S, Martinelli I. Pregnancy associated deep vein thrombosis in a double homozygous carrier of factor V Leiden and prothrombin G20210A. Thrombosis and Haemostasis. 2005;**94**(6):1329-1330

[17] Tanaka Y, Ueda K, Ozawa T, Kitajima I, Okamura S, Morita M, et al. Mutation study of antithrombin: The roles of disulfide bonds in intracellular accumulation and formation of Russell body–like structures. Journal of Biochemistry. 2005;**137**(3):273-285

[18] Alireza R, Rezaie HG. Anticoagulant and signaling functions of antithrombin. Journal of Thrombosis and Haemostasis. 2020;**18**(12):3142-3153

[19] Dhawan A, Eng C. Is the MTHFR gene mutation associated with thrombosis? Cleveland Clinic Journal of Medicine. 2023;**90**(11):661-663

[20] Shibui Y, Sakai R, Manabe Y, Masuyama T. Comparisons of l-cysteine and d-cysteine toxicity in 4-week repeated-dose toxicity studies of rats receiving daily oral administration. Journal of Toxicologic Pathology. 2017;**30**(3):217-229

[21] Plaza NC, García-Galbis MR, Martínez-Espinosa RM. Effects of the usage of l-cysteine (l-Cys) on human health. Molecules. 2018;**23**(3):575

[22] Amin AF, Shaaban OM, Bediawy MA. N-acetyl cysteine for treatment of recurrent unexplained pregnancy loss. Reproductive Biomedicine Online. 2008;**17**(5):722-726

[23] Ghulmiyyah L, Sibai B. Maternal mortality from preeclampsia/eclampsia. Seminars in Perinatology. 2012;**36**(1):56-59

[24] Fiore G, Capasso A. Effects of vitamin E and C on placental oxidative stress: In vitro evidence for the potential therapeutic or prophylactic treatment of preeclampsia. Medicinal Chemistry. 2008;**4**(6):526-530

[25] Motawei SM, Gouda HE, El-Mansoury AM. Effect of N-acetyl cysteine supplementation on blood Lead levels in pregnant women suffering from pre-eclampsia. International Journal of Gynaecology and Obstetrics. 2016;**135**(2):226-227

[26] Bedaiwy MA, Al Inany ARH, Falcone T. N-acetyl cystein improves pregnancy rate in long standing unexplained infertility: A novel mechanism of ovulation induction. Fertility and Sterility. 2004;**82**(2):S228

[27] Kaltsas A, Zikopoulos A, Moustakli E, Zachariou A, Tsirka G, Tsiampali C, et al. The silent threat to Women's fertility: Uncovering the devastating effects of oxidative stress. Antioxidants (Basel). 2023;**12**(8):1490

[28] Yang M, Smith BC. Cysteine and methionine oxidation in thrombotic disorders. Current Opinion in Chemical Biology. 2023;**76**:102350

[29] Wang Q et al. Oxidative stress and thrombosis during aging: The roles of oxidative stress in RBCs in venous thrombosis. International Journal of Molecular Sciences. 2020;**21**(12):4259

[30] Yang M et al. Cysteine sulfenylation by CD36 signaling promotes arterial thrombosis in dyslipidemia. Blood Advances. 2020;**4**(18):4494-4507

[31] Xiaoyun F, Cate SA, Dominguez M, Osborn W, Özpola T, Konkle BA,

et al. Cysteine disulfides (Cys-ss-X) as sensitive plasma biomarkers of oxidative stress. Scientific Reports. 2019;**9**(1):115

[32] Passam FH et al. Beta 2 glycoprotein I is a substrate of thiol oxidoreductases. Blood. 16 Sep 2010;**116**(11):1995-1997

[33] Kumar S et al. An allosteric redox switch in domain V of beta(2)-glycoprotein I controls membrane binding and anti-domain I autoantibody recognition. Journal of Biological Chemistry. 2021;**297**(2):100890

[34] Bai J, Jiao F, Salmeron AG, Shi X, Xian M, Huang L, et al. Mapping pregnancy-dependent Sulfhydrome unfolds diverse functions of protein Sulfhydration in human uterine artery. Endocrinology. 2023;**164**(9):bqad107

[35] Schalinske KL, Anne L, Smazal A. Homocysteine imbalance: A pathological metabolic marker. Advances in Nutrition. 2012;**3**(6):755-762

[36] Giuseppe D, Pamela M. A review about biomarkers for the investigation of vascular function and impairment in diabetes mellitus. Vascular Health and Risk Management. 2016;**12**:415-419

[37] Hadi HA, Carr CS, Al SJ. Endothelial dysfunction: Cardiovascular risk factors, therapy and outcome. Vascular Health and Risk Management. 2005;**1**(3):183-198

[38] Barter PJ, Rye K-A. Homocysteine and cardiovascular disease, is HDL the link? Circulation Research. 2006;**99**(6):565-566

[39] Salemi G, Gueli MC, Vitale F, et al. Blood lipids, homocysteine, stress factor and vitamins in clinically stable multiple sclerosis patients. Lipids in Health and Disease. 2010;**9**:19

[40] Zhang S, Yong-Yi B, Luo LM, Xiao WK, Wu HM, Ye P. Association between serum homocysteine and arterial stiffness in elderly: A community-based study. Journal of Geriatric Cardiology. 2014;**11**(1):32-38

[41] Hassan A, Hunt BJ, O'Sullivan M, Bell R, D'Souza R, Jeffery S, et al. Homocysteine is a risk factor for cerebral small vessel disease, acting via endothelial dysfunction. Brain. 2004;**127**(Pt 1):212-219

[42] Pushpakumar S, Kundu S, Sen U. Endothelial dysfunction: The link between homocysteine and hydrogen Sulfide. Current Medicinal Chemistry. 2014;**21**(32):3662-3672

[43] Lai WK, Kan MY. Homocysteine-induced endothelial dysfunction. Annals of Nutrition & Metabolism. 2015;**67**(1):1-12

[44] Russell VL. Cardiovascular disease: Rise, fall, and future prospects. Annual Review of Public Health. 2011;**32**:1-3

[45] Candido R, Zanetti M. Current perspective. Diabetic vascular disease: From endothelial dysfunction to atherosclerosis. Italian Heart Journal. 2005;**6**(9):703-720

[46] Saposnik G, Ray JG, Sheridan P, McQeen M, Lonn E. Homocysteine-lowering therapy and stroke risk, severity and disability: Additional findings from HOPE 2 trial. Stroke. 2009;**40**(4):1365-1372

[47] Humphrey LL, Fu R, Rogers K, Freeman M, Helfand M. Homocysteine level and coronary disease: A systematic review and meta-analysis. Mayo Clinic Proceedings. 2008;**83**(11):1203-1212

[48] Melichar B, Kalabova H, Krcmova L, et al. Serum homocysteine, cholesterol,

α-tocopherol, glycosylated hemoglobin and inflammatory response during therapy with bevacizumab, oxaliplatin, 5-fluorouracil and leucovorin. Anticancer Research. 2009;**29**(11):4813-4820

[49] Seshadri S, Wolf PA, Beiser AS, Selhub J, Rhoda A, Jacques PF, et al. Association of Plasma Total Homocysteine Levels with subclinical brain injury. Archives of Neurology. 2008;**65**(5):642-649

[50] Vidal J-S, Dufouil C, Ducros V, Tzourio C. Homocysteine, folate and cognition in a large community-based sample of elderly people - the 3C Dijon study. Neuroepidemiology. 2008;**30**(4):207-214

[51] Zylberstein DE, Skoog I, Björkelund C, Guo X, Hultén B, Andreasson L-A, et al. Homocysteine levels and lacunar brain infarcts in elderly women: The prospective population study of women in Gothenburg. Journal of the American Geriatrics Society. 2008;**56**(6):1087-1091

[52] David Smith A, Refsum H, Bottiglieri T, Fenech M, Hoosmand B, McCaddon A, et al. Homocysteine and dementia: An international consensus statement. Journal of Alzheimer's Disease. 2018;**62**(2):561-570

[53] Nelen WL, Steegers EA, Eskes TK, et al. Genetic risk factor for unexplained recurrent early pregnancy loss. Lancet. 1997;**350**(9081):861

[54] Merviel P, Cabry R, Lourdel E, Lanta S, Amant C, Copin H, et al. Comparison of two preventive trataments for pacient with reccurent miscarriages carrying C677T methylenetetrahydrofolate reductas: 5-years' experience. The Journal of International Medical Research. 2017;**45**(6):1720-1730

[55] Abdelsalam T, Karkour T, Elbordiny M, Shalaby D, Abouzeid ZS. Thrombophilia gene mutations in relation to recurrent miscarriage. International Journal of Reproduction, Contraception, Obstetrics and Gynecology. 2018;**7**(3):796-800

[56] Lentzt SR, Evan Sadle J. Inhibition of thrombomodulin surface expression and protein C activation by the thrombogenic agent homocysteine. The Journal of Clinical Investigation. 1991;**88**(6):1906-1914

[57] Den Heijer M, Koster T, Blom HJ, Bos GM, Briët E, Reitsma PH, et al. Hyperhomocysteinemia as a risk factor for deep-vein thrombosis. The New England Journal of Medicine. 1996;**334**(12):759-762

[58] Debreceni L. Homocysteine—A risk factor for atherosclerosis. Orvosi Hetilap. 2001;**142**(27):1439-1444

[59] Milosevic-Tosic M, Borota J. Hyperhomocysteinemia—A risk factor for development of occlusive vascular diseases. Medicinski Pregled. 2002;**55**(9-10):385-391

[60] Falcon CR, Cattaneo M, Panzeri D, Martinelli I, Mannucci PM. Highprevalenceofhyperhomocysteinemia in patients with juvenile venous thrombosis. Arteriosclerosis and Thrombosis. 1994;**14**(7):1080-1083

[61] Cristiana F, Nina Z, Elena A. Blood Cell – An overview of studies in Hematology, Homocysteine in Red Blood Cells Metabolism–Pharmacological Approaches. London, UK, London, UK: InTech; 2012. pp. 31-68, ISBN 978-953-51-0753-8

[62] Sato Y, Honda Y, Iwamoto J, Kanoko T, Satoh K. Effect of folate and mecobalamin on hip fractures in patients

with stroke: A randomized controlled trial. Journal of the American Medical Association. 2005;**293**(9):1082-1088

[63] Rhew EY, Lee C, Eksarko P, Dyer AR, Tily H, Spies S, et al. Homocysteine, bone mineral density, and fracture risk over 2 years of follow-up in women with and without systemic lupus erythematosus. The Journal of Rheumatology. 2008;**35**(2):230-236

[64] Green TJ, McMahon JA, Murray Skeaff C, Williams SM, Whiting SJ. Lowering homocysteine with B vitamins has no effect on biomarkers of bone turnover in old persons:2-y randomized controlled trial. The American Journal of Clinical Nutrition. 2007;**85**(2):460-464

[65] Cagnacci A, Bagni B, Zini A, Cannoletta M, Generali M, Volpe A. Relation of folates, vitamin B12 and homocysteine to vertebral bone mineral density change in postmenopausal women. A five-year longitudinal evaluation. Bone. 2008;**42**(2):314-320

[66] Filip A, Filip N, Veliceasa B, Filip C, Alexa O. The relationship between homocysteine and fragility fractures - a systematic review. Annual Research & Review in Biology. 2017;**16**(5):1-8

[67] Filip N, Cojocaru E, Filip A, Veliceasa B, Alexa O. Reactive oxygen species (ROS) in living cells Edited by InTech. In: Chapter 4 Reactive Oxygen Species and Bone Fragility. London, UK, London, UK: InTech; 2018. pp. 49-67

[68] Brattström L, Wilcken DEL. Homocysteine and cardiovascular disease: Cause or effect? The American Journal of Clinical Nutrition. 2000;**72**(2):315-323

[69] Shukla N, Angelini GD, Jeremy JY. The administration of folic acid reduces intravascular oxidative stress in diabetic rabbits. Metabolism. 2008;**57**(6):774-781

[70] Terzic-Avdagic M. Correlation of coronary disease in patients with diabetes mellitus type 2. Medicinski Arhiv. 2009;**63**(4):191-193

[71] Snoki K, Iwase M, Sasaki N, Ohdo S, Higuchi S, Matsuyama N, et al. Relations of lysophosphatidylcholine in low-density lipoprotein with serum lipoprotein-associated phospholipase A2, paraoxonase and homocysteine thiolactamase activities in patients with type 2 diabetes mellitus. Diabetes Research and Clinical Practice. 2009;**86**(2):117-123

[72] Sen U, Rodriguez WE, Tyagi N, Kumar M, Kundu S, Tyagi SC. Ciglitazone a PPAR γ agonist, ameliorate diabetic nephropathy in part through homocysteine clearance. American Journal of Physiology. Endocrinology and Metabolism. 2008;**295**(5):E1205-E1212

[73] Jia W, Yuan Q, Liang Y-p, Wang H-m, Han X-q, Yin S-q, et al. Serum metrix metalloproteinase-9 combined with homocysteine, IL-6, TNF-α, CRP, HbA1c and lipid profile in the incipient diabetic nephropathy with or without macrovascular diseases. Journal of Medical Colleges of PLA. 2007;**22**(2):111-114

[74] Friedman AN, Hunsicker LG, Selhub J, Bostom AG. Total plasma homocysteine and arteriosclerotic outcomes in type 2 diabetes with nephropathy. Journal of the American Society of Nephrology. 2005;**16**(11):3397-3402

[75] Mangge H, Becker K, Fuchs D, Gostner JM. Antioxidants, inflammation and cardiovascular disease. World Journal of Cardiology. 2014;**6**(6):462-477

[76] Oudi MEL, Aouni Z, Mazigh C, Khochkar R, Gazoueni E, Haouela H, et al. Homocysteine and

markers of inflammation in acute coronary syndrome, exp. Clinical Cardiology. 2010;**15**(2):e25-e28

[77] Pang X, Liu J, Zhao J, Mao J, Zhang X, Feng L, et al. Homocysteine induces the expression of C - reactive protein via NMDAr-ROS-MAPK-NF-κB signal pathway in rat vascular smooth muscle cells. Atherosclerosis. 2014;**236**(1):73-81

[78] Sibrian-Vazquez M, Escobedo JO, Lim S, Samoei GK, Strongin RM. Homocystamides promote free-radical and oxidative damage to proteins. Proceedings of the National Academy of Sciences of the United States of America. 2010;**107**(2):551-554

[79] Papatheodorou L, Weiss N. Vascular oxidant stress and inflammation in hyperhomocysteinemia. Antioxidants & Redox Signaling. 2007;**9**(11):1941-1958

[80] Zou CG, Banerjee R. Homocysteine and redox signaling. Antioxidants & Redox Signaling. 2005;**7**(5-6):547-559

[81] Chuang CH, Lee YY, Sheu BF, Hsiao CT, Loke SS, Chen JC, et al. Homocysteine and C-reactive protein as useful surrogate markers for evaluating CKD risk in adult. Kidney & Blood Pressure Research. 2013;**37**(4-5):402-413

[82] Weiss N, Heydrick SJ, Postea O, Keller C, Keaney JF Jr, Loscalzo J. Influence of hyperhomocysteinemia on the cellular redox state--impact on homocysteine-induced endothelial dysfunction. Clinical Chemistry and Laboratory Medicine. 2003;**41**(11):1455-1461

[83] Filip C, Albu E, Zamosteanu N, Irina MJ, Silion M. Hyperhomocysteinemia's effect on antioxidant capacity on rats. Central European Journal of Medicine. 2010;**5**(5):620-626

[84] Albu E, Filip C, Zamosteanu N, Jaba IM, Linic IS, Sosa I. Hyperhomocysteinemia is an indicator of oxidant stress. Medical Hypotheses. 2012;**78**(4):554-555

[85] Mittal M, Siddiqui MR, Tran K, Reddy SP, Malik AB. Reactive oxygen species in inflammation and tissue injurry. Antioxidants & Redox Signaling. 2014;**20**(7):1126-1167

[86] Wang R, Wang I, Mu N, Lou X, Li W, Chen Y, et al. Activation of NLRP3 inflammasomes contributes to hyperhomocysteinemia-aggravated inflammation and atherosclerosis in apoE-deficient mice. Laboratory Investigation. 2017;**97**(8):922-934

[87] Xi H, Zhang Y, Xu Y, Yang WY, Jiang SX, Cheng X, et al. Caspase-1 inflammasome activation mediates homocystein induced pyro-apoptosis in endothelial cells. Circulation Research. 2016;**118**(10):1525-1539

[88] Shastry S, James LR. Homocysteine-induced macrophage inflammatory protein-2 production by glomerular mesangial cells is mediated by PI3 kinase and p38 MAPK. Journal of Inflammation. 2009;**6**:27

[89] Zhang X, Chen S, Li L, Wang Q, Le W. Folic acid protects motor neurons against the increased homocysteine, inflammation and apoptosis in SOD1G93A transgenic mice. Neuropharmacology. 2008;**54**(7):1112-1119

[90] Hansson GK. Inflammation, atherosclerosis and coronary disease. The New England Journal of Medicine. 2005;**352**(16):1685-1695

[91] Libby P, Theroux P. Pathophysiology of coronary artery disease. Circulation. 2005;**111**(25):3481-3488

[92] Aparna P, Betigeri AM, Pasupath P. Homocysteine and oxidative stress markers and inflammation in patients with coronary artery disease. International Journal of Biological & Medical Research. 2010;**1**(4):125-129

[93] Gori AM, Sofi F, Marcucci R, Abbate R. Association between homocysteine, vitamin B6 concentrations and inflammation. Clinical Chemistry and Laboratory Medicine. 2007;**45**(12):1728-1736

[94] Ganguly P, Alam SF. Role of homocysteine in the development of cardiovascular disease. Nutrition Journal. 2015;**14**:6

[95] Li T, Chen Y, Li J, Yang X, Zhang H, Qin X, et al. Serum homocysteine concentration is significantly associated with inflammatory/immune factors. PLoS One. 2015;**10**(9):e0138099

[96] Schieber M, Chandel NS. ROS function in redox signaling and oxydative stress. Current Biology. 2014;**24**(10):R453-R462

[97] Murphy MP. Mitochondrial thiols in antioxidant protection and redox signaling: Distinct roles for glutathionylation and other thiol modifications. Antioxidants & Redox Signaling. 2012;**16**(6):476-495

[98] Forman HJ, Maiorino M, Ursini F. Signaling function of reactive oxygen species. Biochemistry. 2010;**49**(5):835-842

[99] Czech MP. Differential effects of sulfhydryl reagents on activation and deactivation of the fat cell hexose transport system. The Journal of Biological Chemistry. 1976;**251**(4):1164-1170

[100] Mukherjee SP, Lane RH, Lynn WS. Endogenous hydrogen peroxide and peroxidative metabolism in adipocytes in response to insulin and sulfhydryl reagents. Biochemical Pharmacology. 1978;**27**(22):2589-2594

[101] Mukherjee SP, Mukherjee C. Similar activities of nerve growth factor and its homologue proinsulin in intracellular hydrogen peroxide production and metabolism in adipocytes. Trans-membrane signaling relative to insulin-mimicking cellular effects. Biochemical Pharmacology. 1982;**31**(20):3163-3172

[102] Winterbourn CC, Hampton MB. Thiol chemistry and specificity in redox signaling. Free Radical Biology & Medicine. 2008;**45**(5):549-561

[103] Carty NC, Xu J, Kurup P, Brouillette J, Goebel-Goody SM, Austin DR, et al. The tyrosine phosphatase STEP: Implications in schizophrenia and the molecular mechanism underlying antipsychotic medications. Translational Psychiatry. 2012;**2**(7):e137

[104] Finkel T. Signal transduction by reactive oxygen species. The Journal of Cell Biology. 2011;**194**(1):7-15

[105] Louro RO, Diaz-Moreno I. Redox Proteins in Super Complexes and Signalosomes. Boca Raton, London, New York: CRC Press, Taylor & Francis Group; 2016. p. 299

[106] Kobayashi Y, Ito K, Kanda A, Tomoda K, Miller-Larsson A, Barnes PJ, et al. Protein tyrosine phosphatase PTP-RR regulates corticosteroid sensitivity. Respiratory Research. 2016;**17**:30

[107] Marino SM, Gladyshev VN, Marino SM, Gladyshev VN. Cysteine function governs its conservation and degeneration and restricts its utilization on protein surface. Journal of Molecular Biology. 2010;**404**(5):902-916

[108] Bahorun T, Soobratte MA, Luximon-Ramma V, Aruoma OI. Free radicals and antioxidants in cardiovascular health and disease. Internet Journal of Medical Update. 2006;**1**(2):25-41

[109] Novo E, Parola M. Redox mechanisms in hepatic chronic wound healing and fibrogenesis. Fibrogenesis & Tissue Repair. 2008;**1**(1):5

[110] Martindale JL, Holbrook NJ. Cellular response to oxidative stress: Signaling for suicide and survival. Journal of Cellular Physiology. 2002;**192**(1):1-15

[111] Powers SK, Duarte J, Kavazis AN, Talbert EE. Reactive oxygen species are signaling molecules for skeletal muscle adaptation. Experimental Physiology. 2010;**95**(1):1-9

[112] Sun JP, Zhang ZY, Wang WQ. An overview of the protein tyrosine phosphatase super family. Current Topics Medicinal Chemistry. 2003;**3**(7):739-748

[113] Rhee S, Cell signaling. H_2O_2, a necessary evil for cell signaling. Science. 2006;**312**(5782):1882-1883

[114] Tsen CM, Hsieh CC, Yen CH, Lau YT. Homocysteine altered ROS generation and NO accumulation in endothelial cells. The Chinese Journal of Physiology. 2003;**46**(3):129-136

[115] Zammiti W, Mtiraoui N, Mahjoub T. Lack of consistent association between endothelial nitric oxide synthase gene polymorphisms, homocysteine levels and recurrent pregnancy loss in Tunisian women. American Journal of Reproductive Immunology. 2008;**59**(2):139-145

[116] Erol A, Çınar MG, Can C, Olukman M, Ülker S, Koşay S. Effect of homocysteine on nitric oxide production in coronary microvascular endothelial cells. Endothelium. 2007;**14**(3):157-161

[117] Dai C, Fei Y, Li J, Shi Y, Yang X. A novel review of homocysteine and pregnancy complications. BioMed Research International. 2021;**2021**:6652231

[118] Cikot RJLM, Steegers-Theunissen RP, Thomas CM, de Boo TM, Merkus HM, Steegers EA. Longitudinal vitamin and homocysteine levels in normal pregnancy. The British Journal of Nutrition. 2001;**85**(1):49-58

[119] Walker MC, Smith GN, Perkins SL, Keely EJ, Garner PR. Changes in homocysteine levels during normal pregnancy. American Journal of Obstetrics and Gynecology. 1999;**180**(3 Pt 1):660-664

[120] Yang Y, Jiang H, Tang A, Xiang Z. Changes of serum homocysteine levels during pregnancy and the establishment of reference intervals in pregnant Chinese women. Clinica Chimica Acta. 2019;**489**:1-4

[121] de Vries JIP, van Pampus MG, Hague WM, Bezemer PD, Joosten JH, on behalf of Fruit Investigators. Low-molecular-weight heparin added to aspirin in the prevention of recurrent early-onset pre-eclampsia in women with inheritable thrombophilia: The FRUIT-RCT. Journal of Thrombosis and Haemostasis. 2012;**10**(1):64-72

[122] Di Simone N, Maggiano N, Caliandro D, et al. Homocysteine induces trophoblast cell death with apoptotic features. Biology of Reproduction. 2003;**69**(4):1129-1134

[123] Acharya N. Homocysteinemia: A rare cause of recurrent pregnancy loss coexisting with deep vein thrombosis. Journal of South Asian Federation of Obstetrics and Gynaecology. 2020;**12**(5):328-330

[124] Langman LJ, Ray JG, Evrovski J, Yeo E, Cole DE. Hyperhomocyst(e)inemia and the increased risk of venous thromboembolism: More evidence from a case-control study. Archives of Internal Medicine. 2000;**160**(7):961-964

[125] Kamel H, Navi BB, Sriram N, Hovsepian DA, Devereux RB, Elkind MSV. Risk of a thrombotic event after the 6-week postpartum period. The New England Journal of Medicine. 2014;**370**(14):1307-1315

Chapter 5

New Insights into the Roles of Cysteine and Homocysteine in Pathological Processes

Nina Filip, Alin Constantin Pinzariu, Minela Aida Maranduca, Diana Zamosteanu and Ionela Lacramioara Serban

Abstract

Both cysteine and homocysteine are sulfur-containing amino acids with distinct roles in cellular processes. This chapter explores novel perspectives on the roles of cysteine and homocysteine in pathological processes, delving into their intricate involvement in various disease pathways. Additionally, the chapter elucidates the regulatory mechanisms governing homocysteine metabolism and its implications for a range of pathological conditions, including cardiovascular diseases and neurodegenerative disorders. By synthesizing recent research findings, this chapter aims to provide fresh insights into the nuanced interplay among cysteine, homocysteine, and disease progression. The exploration of these sulfur-containing amino acids opens avenues for understanding pathophysiological mechanisms and suggests potential targets for therapeutic interventions.

Keywords: cysteine, homocysteine, cardiovascular disease, diabetes mellitus, amino acids

1. Introduction

Cysteine, an amino acid of particular importance in cell biology, serves as an essential component in the construction of the fundamental building blocks of life—proteins. With a distinctive chemical structure, characterized by the presence of the thiol group (–SH) in its side chain, cysteine holds crucial roles in stabilizing protein structure and maintaining oxidative balance in the body's cells [1, 2]. Although, usually, the body can synthesize cysteine, there are circumstances in which its necessity becomes conditionally essential, justifying the importance of adequate intake through food or, in certain cases, supplementation. The various aspects of cysteine, from its contribution to the formation of disulfide bonds into proteins to its antioxidative capacity and involvement in cellular detoxification processes, provide clear insight into the importance of cysteine in maintaining cellular health and essential physiological functions [3].

Homocysteine, an organic compound derived from the essential amino acid methionine, has become the subject of considerable interest in the scientific and

IntechOpen

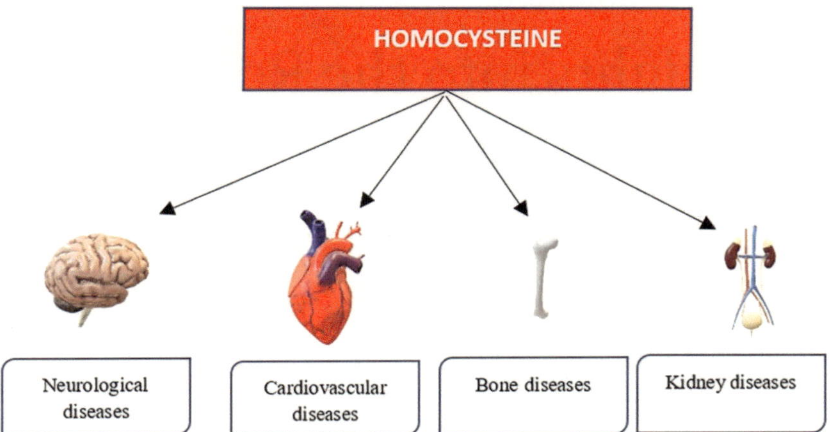

Figure 1.
Homocysteine associated diseases.

medical field due to its link to various conditions and diseases. Normally, methionine, an amino acid obtained from food, is metabolized in the body to produce homocysteine, and its levels are strictly regulated [4, 5]. However, when this balance is disrupted, homocysteine can reach high levels, associated with an increased risk for cardiovascular disease, neurological disease, and others (**Figure 1**) [6].

Homocysteine metabolism is closely related to adequate intake of B vitamins, especially B6, B9 (folic acid), and B12. Deficiencies in these vitamins can lead to the accumulation of homocysteine in the blood. Genetic factors can also influence homocysteine levels [7]. Studies have shown that an increased level of homocysteine in the blood may be an independent risk factor for cardiovascular disease, as it can contribute to damage to the vascular wall and promote blood clots [8, 9]. It has also been suggested that homocysteine may play a role in inflammatory and degenerative processes in the nervous system, contributing to neurodegenerative diseases, such as Alzheimer's disease or Parkinson's disease [10, 11]. Monitoring homocysteine levels and adopting a healthy lifestyle that includes a balanced diet and appropriate B vitamin supplements can help keep these levels within normal limits and reduce the risk associated with elevated homocysteine levels [12].

In this presentation, we explore the nature and roles of cysteine and homocysteine in the body, analyzing their implications for health and their connections to various pathologies. Understanding these molecules provides a significant perspective on how nutritional and genetic factors can influence cardiovascular health and neurological functions, prompting research and interventions for the prevention and management of complications associated with elevated levels of homocysteine.

2. Cysteine and homocysteine metabolism

Cysteine and homocysteine are amino acids involved in important metabolic pathways within the body. Here's an overview of their metabolism.

Homocysteine, a sulfur-containing amino acid, is produced in the metabolic process of methionine, an essential amino acid present in animal-derived proteins [13]. This compound follows two primary metabolic pathways: remethylation, leading to the regeneration of methionine, and transsulfuration, resulting in its conversion to cysteine. Methionine serves as the primary supplier of methyl radicals in the body, delivered in the form of S-adenosylmethionine (SAM), which is subsequently transformed into S-adenosylhomocysteine (SAH). Homocysteine is produced through the breakdown of SAH, a process catalyzed by S-adenosylhomocysteine hydrolase (SAHH; EC 3.3.1.1). Enzymatic processes play a crucial role in the metabolism of homocysteine, a sulfur-containing amino acid. These processes involve various enzymes, such as methionine synthase (MS; EC 2.1.1.13) and N5,10-methylenetetrahydrofolate reductase (MTHFR; EC 1.5.1.20), which catalyze important reactions in the conversion of homocysteine to methionine. Additionally, the activity of these enzymes is dependent on cofactors such as cobalamin and N5-methyltetrahydrofolate. The genetic association between the genes encoding methionine synthase and N5,10-methylenetetrahydrofolate reductase suggests a close relationship between these enzymes and the tetrahydrofolate metabolism (**Figure 2**) [14–17].

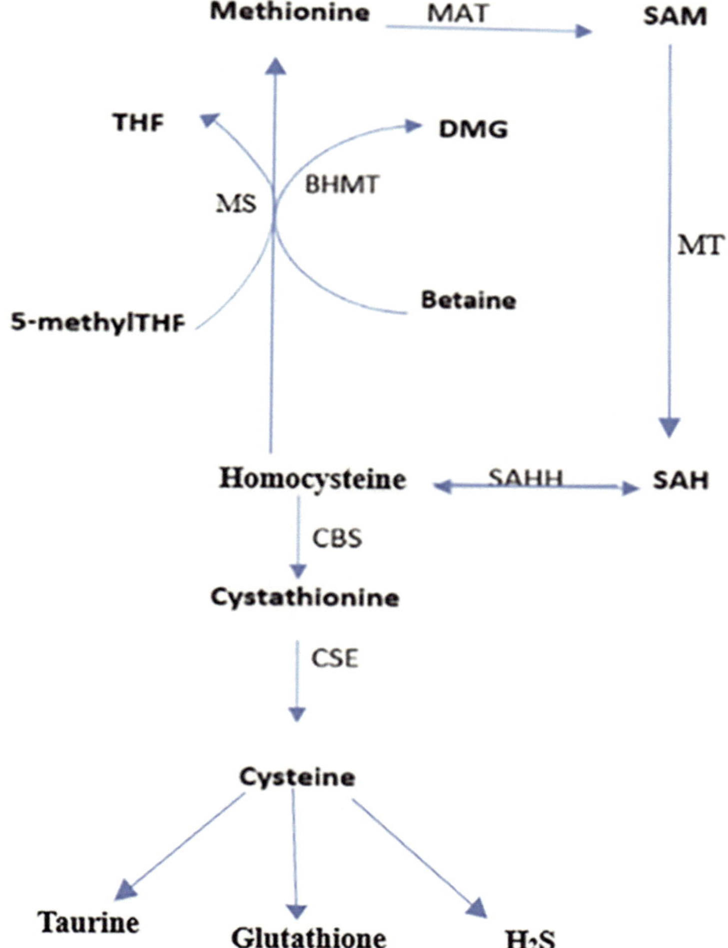

Figure 2.
Homocysteine metabolism. Methionine adenosyltransferase (MAT); methyltransferase (MT); S-adenosylhomocysteine hydrolase (SAHH); betaine-homocysteine S-methyl transferase (BHMT); cystathionine β-synthase (CBS); cystathionine γ-lyase (CSE); methionine synthase (MS). Adapted from Filip et al. [14].

The remethylation of homocysteine is an important biochemical process that plays a crucial role in maintaining optimal health and functioning in the body. There are multiple pathways involved in the remethylation of homocysteine, including the betaine pathway and the methionine synthase pathway. The betaine pathway utilizes betaine, derived from choline, and is facilitated by the enzyme betaine-homocysteine S-methyltransferase (BHMT; EC 2.1.1.5) [14, 18, 19]. This pathway is primarily active in the liver and kidneys. The methionine synthase pathway, on the other hand, is present in all tissues of the body. One-carbon metabolism is a network of interrelated biochemical reactions that includes the remethylation of homocysteine. One-carbon metabolism is essential for DNA methylation and synthesis, with methionine serving as the main methyl group donor. Not only does the remethylation of homocysteine contribute to methyl group synthesis, but it also serves as a means to prevent the intracellular accumulation of S-adenosylhomocysteine and an increase in plasma total homocysteine levels [14, 20–22]. Accumulation of elevated levels of homocysteine and adenosine at the cellular level has been demonstrated to completely inhibit all methylation reactions [23–25]. The conversion of homocysteine to cysteine occurs through cystathionine.

Cystathionine β-synthase (CBS; EC 4.2.1.22) and cystathionine γ-lyase (CSE; EC 4.4.1.1) are enzymes involved in the synthesis of cysteine (**Figure 2**). Cystathionine β-synthase and cystathionine γ-lyase play crucial roles in the metabolism of cysteine, a non-essential amino acid that is important for various physiological processes in the body. These enzymes facilitate the conversion of homocysteine and serine into cystathionine, which is then further processed to produce cysteine and other essential molecules. The synthesis of cysteine relies on the activity of cystathionine β-synthase and cystathionine γ-lyase enzymes. These enzymes work together to catalyze the reactions necessary for cysteine synthesis. They both require pyridoxal-5-phosphate, the active form of vitamin B6, as a cofactor for their enzymatic activity [26–28]. Moreover, alterations in the activity or concentration of cystathionine β-synthase and cystathionine γ-lyase have been associated with various pathological conditions and diseases, including diabetes, cancer, AIDS, neurodegenerative disorders, and liver diseases [29–31]. Furthermore, the enzymes cystathionine β-synthase and cystathionine γ-lyase are also involved in maintaining glutathione metabolism and transport [31, 32]. These enzymes contribute to cellular reactions, such as antioxidant defense, drug detoxification, and cell signaling [33].

Activation of the transsulfuration pathway can promote the production of H_2S [34]. Certainly, cysteine metabolism is intricately linked to methionine metabolism. The process of synthesizing cysteine from methionine comprises four sequential steps. The first two stages are integrated into the methionine cycle, while the subsequent two phases are integral components of the transsulfuration pathway. This interconnected series of events underscores the dynamic relationship between cysteine and methionine within cellular metabolism. The process commences with methionine, leading to the formation of S-adenosylmethionine. Subsequently, SAM contributes to the synthesis of homocysteine. In the initial step of the transsulfuration pathway, cystathionine-β-synthase enzymatically couples homocysteine with serine, resulting in the generation of cystathionine. Subsequent to this, in the subsequent reaction, cystathionine undergoes cleavage catalyzed by cystathionin-γ-lyase, yielding cysteine. This sequence of reactions elucidates the transformation of methionine into cysteine through the transsulfuration pathway [35]. To finalize the synthesis of the GSH tripeptide from cysteine, two additional reactions are essential (**Figure 3**).

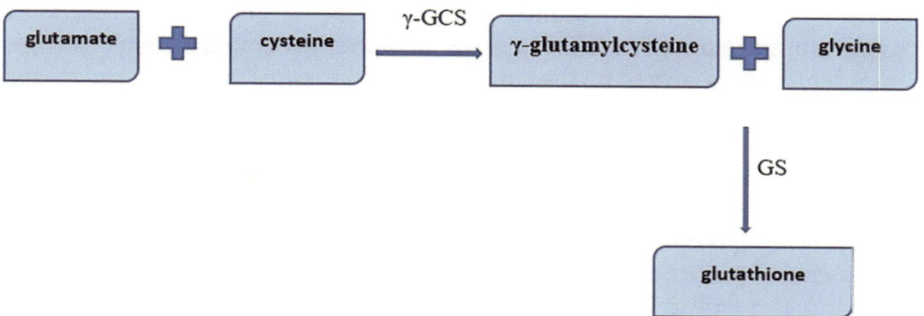

Figure 3.
Scheme of glutathione (GSH) biosynthesis. Adapted from Potega et al. [36]. GS = glutathione synthetase; γ-GCS = γ-glutamylcysteine synthetase.

Initially, the enzyme γ-glutamyl cysteine synthetase (γ-GCS) facilitates the combination of cysteine with glutamate, leading to the creation of γ-glutamylcysteine. Subsequently, in the second reaction catalyzed by the enzyme glutathione synthetase (GS), GSH is produced from γ-glutamylcysteine and glycine [36]. GSH, also known as glutathione, possesses the ability to undergo oxidation, thereby contributing to the maintenance of redox homeostasis. In a reducing environment, such as in the presence of reactive oxygen species (ROS), a disulfide bridge forms between two GSH molecules [37]. This linkage results in the formation of glutathione disulfide (GSSG). The capacity of GSH to undergo redox transformations is crucial for its involvement in cellular antioxidant defense mechanisms. Cysteine synthesis and homocysteine metabolism intersect in the transsulfuration pathway. This pathway can influence the balance between cysteine and homocysteine levels.

3. Cysteine, homocysteine, and mitochondrial dysfunction

The interplay among cysteine, homocysteine, and mitochondrial dysfunction is intricate, involving diverse biochemical processes. Cysteine and homocysteine, being sulfur-containing amino acids, possess the potential to impact mitochondrial function through various metabolic pathways. The complexity of this relationship stems from the multifaceted influence these amino acids exert on mitochondrial processes [38].

Hyperhomocysteinemia induces DNA fragmentation via poly (ADP-ribose) polymerase cleavage, an enzymatic activity effectively suppressed by melatonin. The impact of homocysteine on the function of thioretinaco ozonide within mitochondria and endoplasmic reticulum leads to excitotoxicity, oxidative stress, endothelial dysfunction, and inflammation [39]. Melatonin reduces oxidative damage in the mitochondria of cultured mouse liver cells by diminishing oxygen consumption, lowering membrane potential, and decreasing superoxide production. Matrix metalloproteinase-9 activation in hyperhomocysteinemic rats is triggered by the opening of the mitochondrial permeability transition pore and antagonism of the N-methyl-D-aspartate receptor-1 (NMDA-R1), resulting in myocyte dysfunction due to elevated calcium overload and oxidative stress [40]. The homocysteine-inducible, endoplasmic reticulum (ER) stress-inducible, ubiquitin-like domain member 2 (HERPUD1) in cultured HELA cells facilitate cytoprotective effects against oxidative stress by suppressing the inositol 1,4,5-triphosphate receptor and reducing the transfer of calcium ions from the endoplasmic reticulum to mitochondria [41]. Homocysteine serves as a precursor to hydrogen sulfide (H_2S), a gasotransmitter generated through the transsulfuration process catalyzed by the enzyme's cystathionine β-synthase and cystathionine γ-lyase. Recently recognized as a novel mediator in cardiovascular homeostasis, H_2S functions as a potent vasodilator with diverse roles [42]. H_2S is primarily generated from L-cysteine, obtained through dietary intake, extraction from endogenous proteins, or endogenous synthesis via the transsulfuration of serine by L-methionine, a process occurring within the transsulfuration pathway. Significantly, L-cysteine serves as a crucial precursor for the biosynthesis of glutathione (GSH), and its initiation into this pathway is a limiting factor [43]. Consequently, mechanisms that modulate the availability of L-cysteine are likely to impact both the H_2S production pathway and GSH biosynthesis, as evidenced by metabolomics and transcriptomics studies conducted in diverse models of mitochondrial dysfunction. As an illustration, investigations into the impact of 1-methyl-4-phenylpyridinium (MPP+), known for its neurotoxic effects mediated by various mechanisms, including complex I (CI) inhibition, demonstrated elevated glutathione (GSH) levels coupled with the upregulation of ATF-4 and the transsulfuration enzymes CTH and CBS. In cellular stress induced by MPP+, the knockdown of ATF-4 or CTH resulted in a reduction of GSH levels [43, 44].

4. The roles of cysteine and homocysteine in pathological processes

4.1 Cysteine and homocysteine in cardiovascular disease

Cardiovascular disease (CVD) stands as the leading cause of global mortality, impacting both developed and developing nations. Numerous studies have explored the relationship between elevated homocysteine (Hcy) levels and the onset of various vascular conditions [45]. Some researchers argue that maintaining a plasma Hcy concentration below 10 μmol/L is essential to mitigate the heightened risk of CVD and ischemic heart disease [46]. A meta-analysis has revealed a positive correlation, indicating that a 25% increase in plasma Hcy levels corresponds to a 10% higher risk of CVD and a 20% higher risk of stroke. Furthermore, another meta-analysis highlighted a 16% reduction in coronary heart disease when serum Hcy levels decreased by 3 μmol/L. Additionally, a 5 μmol/L elevation in plasma Hcy was associated with a 1.6–1.8 times increased relative risk of coronary heart disease [47]. Further investigations have underscored the significant impact of homocysteine on mortality and heart disease [48]. Hcy is implicated in cardiovascular disease through diverse mechanisms, including the heightened proliferation of muscle cells leading to vessel constriction, modification of blood coagulation properties, oxidative damage to the vascular endothelium, and harm to arterial walls [49].

Moreover, Hcy has demonstrated a positive correlation with both diastolic and systolic blood pressures. For example, a 5 μmol/L elevation in Hcy concentration has been correlated with a 0.5 mmHg rise in diastolic blood pressure and a 0.7 mmHg increase in systolic blood pressure. In the case of women, the relationship between Hcy and blood pressure was more pronounced, resulting in a 0.7 mmHg increase in diastolic blood pressure and a 1.2 mmHg increase in systolic blood pressure [50–52].

Epidemiological research has uncovered a U-shaped correlation between cardiovascular diseases and total cysteine (tCys) after adjusting for other risk factors and homocysteine [52]. Van den Brandhof and colleagues [53], in their study, did not observe a significant association between tCys and the risk of coronary heart disease. Similarly, a study within the Hordaland Homocysteine Study cohort explored the link between tCys and the risk of mortality and cardiovascular disease. The findings of this study concluded that tCys was not connected to all-cause cardiovascular disease or non-cardiovascular disease mortality [54].

Subsequent research endeavored to evaluate the association between total homocysteine (tHcy) and cardiovascular disease, along with mortality and morbidity unrelated to CVD. The findings suggested that tHcy functions as a biomarker for both CVD and non-CVD-related mortality and morbidity [33, 54, 55].

4.2 Cysteine and homocysteine in kidney diseases

As evidenced by several studies, the kidneys play a pivotal role in the regulation of the methionine cycle, exerting a direct influence on alterations in the concentrations of homocysteine, cysteine, S-adenosylmethionine, S-adenosylhomocysteine, and the SAH/SAM ratio [56, 57].

Despite not fully comprehending the mechanisms underlying these disorders in chronic kidney disease (CKD), the robust association among homocysteine, cysteine, S-adenosylmethionine, S-adenosylhomocysteine, and plasma creatinine levels, as well as the glomerular filtration rate (GFR), indicated that these compounds were excreted in the urine. The presence of enzymes in the trans-sulfonation and remethylation pathways in human kidney tissue implies the potential involvement of renal metabolism in these processes [58].

Alterations in the concentrations of Cys and Hcy in blood plasma serve as early indicators, implying that there is initially a primary disruption in the aminothiol metabolism. This disturbance persists until a more pronounced decrease in glomerular filtration rate occurs, along with a disruption in the elimination of these compounds [58].

Krugovla et al. [59] reported that individuals at stages III–V of chronic kidney disease exhibit significantly reduced urine levels of S-adenosylmethionine and SAM/S-adenosylhomocysteine ratio, as well as a decreased cysteine/homocysteine ratio in blood plasma in comparison to patients at stage II of CKD. Notably, the levels of urine SAM serve as a discriminative factor, enabling the differentiation between patients with mildly decreased kidney function and those with moderate to severe renal impairment. They demonstrate that urine SAM is a potent biomarker for monitoring renal function decline at early CKD stages [59].

Shih et al. [60] indicated that a high homocysteine level can be an independent risk factor for CKD in the middle-aged and elderly populations in Taiwan.

Homocysteine can contribute to ROS production by activating NADPH oxidase. These generated ROS disrupt normal cellular function, thereby contributing to kidney disease. Studies have indeed demonstrated that hyperhomocysteinemia compromises microvascular function in various organs, including the microvascular system within the kidneys [60].

S-adenosyl-L-methionine, derived from methionine, serves as the substrate for SAM-dependent methyltransferases, which catalyze the transfer of its methyl group. SAM plays a crucial role, accounting for over 90% of the methylation processes involving nucleic acids, proteins, and lipids. The reaction also produces S-adenosylhomocysteine, which, in turn, acts as an inhibitor of the methyltransferase reaction. In conditions of hyperhomocysteinemia, there is an increase in intracellular SAH, leading to the inhibition of SAM-dependent methyl transfer. High SAH levels particularly affect DNA methyltransferases, impacting the methylation status of certain genes, including the hTERT gene, which has implications for kidney function [60].

4.3 Cysteine and homocysteine in diabetes

Cysteine and homocysteine are two sulfur-containing amino acids that have been implicated in the pathogenesis of diabetes. Previous studies have shown that patients with diabetes tend to have higher levels of homocysteine compared to the general population [61].

Diabetic retinopathy is a severe microvascular complication that affects individuals with diabetes, leading to progressive damage to the retina and potentially causing vision loss. Several studies have suggested that homocysteine, a sulfur-containing amino acid, plays a significant role in the development and progression of diabetic retinopathy [61, 62]. Elevated levels of homocysteine have been linked to increased oxidative stress, mitochondrial damage, and activation of various pathways associated with the development and progression of diabetic retinopathy. Multiple mechanisms have been proposed to explain the involvement of homocysteine in diabetic retinopathy. One proposed mechanism is the activation of matrix metalloproteinase-9 by homocysteine, which leads to disruption of the blood-retinal barrier and increased vascular permeability in the retina [61]. Another proposed mechanism involves the inhibition of the transcriptional activity of nuclear factor erythroid 2-related factor 2, a key regulator of antioxidant defense systems, by elevated levels of homocysteine. This inhibition compromises the defense system's ability to counteract increased oxidative stress, leading to further damage to retinal cells. Furthermore, it has been observed that hyperhomocysteinemia is often associated with endothelial dysfunction and inflammation, both of which are known to contribute to the development and progression of diabetic retinopathy. Several studies have investigated the potential therapeutic implications of reducing homocysteine levels in diabetic retinopathy.

Furthermore, high homocysteine levels can trigger inflammation, endoplasmic reticulum, stress, and other pathways, intensifying mitochondrial damage. The biosynthesis of S-adenosylmethionine during homocysteine metabolism from methionine further activates DNA methyltransferases, leading to alterations in the DNA methylation status of various genes implicated in mitochondrial homeostasis [63].

Wijekoon et al. [64] provided experimental evidence showcasing the direct impact of insulin and counter-regulatory hormones on the modulation of cystathionine β-synthase and betaine homocysteine methyltransferase.

Their investigations revealed that insulin exerts a regulatory influence on cystathionine β-synthase and BHMT, contributing to the intricate metabolic changes associated with diabetes mellitus. Additionally, counter-regulatory hormones were found to play a pivotal role in this regulatory network, further shaping the activities of these enzymes. The interplay between insulin and counter-regulatory hormones emerges as a crucial factor in the dysregulation of homocysteine metabolism seen in both Type 1 and Type 2 diabetes. The meta-analysis of Wang et al. [65] indicated elevated blood homocysteine levels in individuals with Type 2 diabetes mellitus (T2DM), particularly among those with diabetic nephropathy (DN) and diabetic retinopathy (DR), but the precise role of homocysteine in the development of T2DM and its associated complications remains ambiguous.

Homocysteine has been identified as closely associated with the development of diabetes mellitus. Homocysteine alone is insufficient as a predictor of diabetes mellitus. Li et al. [66] determined the relationship between the main metabolites involved in the Hcy metabolic pathway and DM. Their results revealed the detection of 13 metabolites associated with the Hcy metabolic pathway in the samples. In the DM group, the levels of Hcy, cysteine, taurine, pyridoxamine, methionine, and choline exhibited a significant increase. Notably, Hcy, choline, cystathionine, methionine, and taurine played a significant role in the probabilistic principal component analysis model. Hcy, taurine, methionine, and choline emerge as potential risk factors for diagnosing diabetes and hold promise for assessing the severity of the condition [66].

A deficiency in L-serine can contribute to elevated homocysteine levels through two mechanisms. Firstly, there is a reduced supply of N5-CH3-THF for the methylation of homocysteine, attributable to an adaptive increase in L-serine synthesis from glycine. Secondly, there is an impaired synthesis of cystathionine from L-serine and homocysteine by cystathionine β-synthase, leading to a subsequent reduction in the drainage of homocysteine from the methionine cycle to the transsulfuration pathway. This possibility is substantiated by the occurrence of hyperhomocysteinemia in humans exhibiting cystathionine β-synthase deficiency [67, 68]. Rehman et al. [33] reported decreased levels of cysteine concomitant with increased homocysteine levels in diabetes.

4.4 Cysteine and homocysteine in neurological disorders

Homocysteine is transported into the brain, and the brain has a limited capacity for Hcy metabolism. Brain tissues employ various mechanisms to reduce homocysteine levels. These mechanisms include efficient recycling through the vitamin B12-dependent methionine synthase, which is the sole enzyme in the brain responsible for converting Hcy into methionine. Additionally, catabolism occurs through cystathionine beta-synthase, leading to the formation of cystathionine—a non-noxious product that further converts into cysteine. Moreover, the brain regulates Hcy levels by exporting them to the external circulation [69].

In the brain, the accumulation of homocysteine is correlated with increased total homocysteine and S-adenosylmethionine levels in cerebrospinal fluid. Induced elevation of Hcy has been shown to induce dysfunction in endothelial and astrocytic cells, leading to altered neuronal function. Elevated Hcy levels result in heightened excitatory glutamatergic neurotransmission in various brain regions, contributing to neuronal damage. In summary, Hcy induces redox imbalance, increases oxidative stress, and triggers the production of free radicals in various cell types, including endothelial, glial, and neuronal cells, thereby contributing to several neurological disorders [70].

Both cysteine and homocysteine metabolism are linked to neurological disorders, and mitochondrial dysfunction is often observed in conditions such as Alzheimer's disease and Parkinson's disease.

5. Identification of current challenges and gaps in understanding cysteine and homocysteine roles

Understanding the roles of cysteine and homocysteine in various physiological processes is a current area of research. Here are some challenges and gaps in understanding these amino acids.

Both amino acids are involved in complex metabolic pathways. Cysteine is derived from methionine via homocysteine, and its metabolism involves complex interactions with various enzymes, cofactors, and regulatory molecules. Understanding the entire metabolic network and the factors that influence it is challenging.

Cysteine plays a crucial role in cellular redox homeostasis, being a key component of glutathione, an important antioxidant. The precise mechanisms by which cysteine influences redox balance in different cellular compartments and its implications for health and disease are not fully elucidated.

Homocysteine is implicated in DNA methylation, which is crucial for gene expression regulation. However, the exact mechanisms by which homocysteine influences epigenetic modifications and the specific genes affected are areas of ongoing investigation.

Elevated homocysteine levels have been associated with neurodegenerative diseases. Understanding the exact mechanisms by which homocysteine contributes to neurological disorders and whether it is a causative factor or a consequence is still an active area of research.

Homocysteine has been linked to cardiovascular diseases, but the relationship is complex. The specific mechanisms by which elevated homocysteine contributes to cardiovascular pathologies and whether interventions to lower homocysteine levels are beneficial remain areas of debate and investigation.

Cysteine and homocysteine metabolism are influenced by dietary factors, including the intake of methionine, vitamins (B6, B12, folate), and other nutrients. The interplay between diet and the regulation of cysteine and homocysteine levels is not fully understood, especially in diverse populations with different dietary patterns.

Both cysteine and homocysteine can interact with various molecules and participate in diverse cellular processes. Understanding the wide-ranging interactions and their implications for cellular function and health requires further investigation.

Identifying reliable biomarkers related to cysteine and homocysteine metabolism for diagnostic and prognostic purposes remains an ongoing challenge. Additionally, determining the clinical significance of alterations in these amino acids and developing targeted therapies is an area of active research.

6. Conclusions

The evolving nature of our understanding of cysteine and homocysteine in pathological processes reflects the dynamic and interdisciplinary nature of scientific research. Scientific research is a continuous process, and new studies are regularly published, providing novel insights into the roles of cysteine and homocysteine in various pathological processes. Ongoing research may uncover previously unknown connections, mechanisms, and functions related to these amino acids. Advances in analytical techniques and technologies contribute to our ability to study cysteine and homocysteine at a more detailed and nuanced level. Techniques such as metabolomics, proteomics, and advanced imaging methods allow researchers to explore these molecules' functions and interactions in greater depth.

The role of homocysteine in epigenetic modifications and gene regulation is an evolving area of interest. Continued research may reveal additional details about how homocysteine influences gene expression and the implications for various diseases.

Ongoing clinical trials and translational research contribute to the translation of basic scientific knowledge into practical applications for diagnosis, treatment, and prevention. As these studies progress, our understanding of the clinical significance of cysteine and homocysteine in specific pathological conditions may evolve. Collaboration between researchers from different fields, such as biochemistry, genetics, nutrition, and clinical medicine, continues to enrich our understanding of cysteine and homocysteine. Interdisciplinary approaches help address complex questions and provide a more comprehensive view of their roles in pathology.

Given the dynamic nature of the scientific inquiry, it is crucial to stay informed about the latest research findings and advancements in the field to appreciate the evolving landscape of our understanding of cysteine and homocysteine in pathological processes.

Conflict of interest

The authors declare no conflict of interest.

Author details

Nina Filip[1*], Alin Constantin Pinzariu[2], Minela Aida Maranduca[2], Diana Zamosteanu[3] and Ionela Lacramioara Serban[2]

1 Faculty of Medicine, Department of Morpho-Functional Sciences (II), Discipline of Biochemistry, “Grigore T. Popa” University of Medicine and Pharmacy, Iasi, Romania

2 Faculty of Medicine, Department of Morpho-Functional Sciences (II), Discipline of Physiology, “Grigore T. Popa” University of Medicine and Pharmacy, Iasi, Romania

3 Pathological Anatomy, “Grigore T. Popa” University of Medicine and Pharmacy, Iasi, Romania

*Address all correspondence to: zamosteanu_nina@yahoo.com

References

[1] Moosmann B, Schindeldecker M, Hajieva P. Cysteine, glutathione and a new genetic code: Biochemical adaptations of the primordial cells that spread into open water and survived biospheric oxygenation. Biological Chemistry. 2020;**401**(2):213-231. DOI: 10.1515/hsz-2019-0232

[2] Ling ZN, Jiang YF, Ru JN, Lu JH, Ding B, Wu J. Amino acid metabolism in health and disease. Signal Transduction and Targeted Therapy. 2023;**8**(1):345. DOI: 10.1038/s41392-023-01569-3

[3] Carballal S, Banerjee R. Overview of cysteine metabolism. In: Redox Chemistry and Biology of Thiols. Cambridge, Massachusetts, United States: Academic Press; 2022. pp. 423-450

[4] Pajares MÁ. Amino acid metabolism and disease. International Journal of Molecular Sciences. 2023;**24**(15):11935. DOI: 10.3390/ijms241511935

[5] Wu C, Duan X, Wang X, Wang L. Advances in the role of epigenetics in homocysteine-related diseases. Epigenomics. 2023;**15**(15):769-795. DOI: 10.2217/epi-2023-0207

[6] McCully KS. The biomedical significance of homocysteine. Journal of Scientific Exploration. 2001;**15**(1):5-20

[7] Sunder-Plassmann G, Födinger M. Genetic determinants of the homocysteine level. Kidney International Supplement. 2003;**84**:S141-S144. DOI: 10.1046/j.1523-1755.63. s84.52. x

[8] Chrysant SG, Chrysant GS. The current status of homocysteine as a risk factor for cardiovascular disease: A mini review. Expert Review of Cardiovascular Therapy. 2018;**16**(8):559-565. DOI: 10.1080/14779072.2018.1497974

[9] Guieu R, Ruf J, Mottola G. Hyperhomocysteinemia and cardiovascular diseases. Annales de biologie clinique. 2022;**80**(1):7

[10] Moretti R, Giuffré M, Caruso P, Gazzin S, Tiribelli C. Homocysteine in neurology: A possible contributing factor to small vessel disease. International Journal of Molecular Sciences. 2021;**22**(4):2051. DOI: 10.3390/ijms22042051

[11] Cordaro M, Siracusa R, Fusco R, Cuzzocrea S, Di Paola R, Impellizzeri D. Involvements of hyperhomocysteinemia in neurological disorders. Metabolites. 2021;**11**(1):37. DOI: 10.3390/metabo11010037

[12] Yuan S et al. Homocysteine, B vitamins, and cardiovascular disease: A Mendelian randomization study. BMC Medicine. 2021;**19**:1-9

[13] Finkelstein JD. Metabolic regulatory properties of S-adenosylmethionine and S-adenosylhomocysteine. Clinical Chemistry and Laboratory Medicine. 2007;**45**(12):1694-1699. DOI: 10.1515/CCLM.2007.341

[14] Filip N, Cojocaru E, Badulescu OV, Clim A, Pinzariu AC, Bordeianu G, et al. SARS-CoV-2 infection: What is currently known about homocysteine involvement? Diagnostics (Basel). 2022;**13**(1):10. DOI: 10.3390/diagnostics13010010

[15] Pascale RM, Simile MM, Calvisi DF, Feo CF, Feo F. S-Adenosylmethionine: From the discovery of its inhibition of tumorigenesis to its use as a

therapeutic agent. Cells. 2022;**11**(3):409. DOI: 10.3390/cells11030409

[16] Mato JM, Martínez-Chantar ML, Lu SC. S-adenosylmethionine metabolism and liver disease. Annals of Hepatology. 2013;**12**(2):183-189

[17] Imbard A, Benoist JF, Esse R, Gupta S, Lebon S, de Vriese AS, et al. High homocysteine induces betaine depletion. Bioscience Reports. 2015;**35**(4):e00222. DOI: 10.1042/BSR20150094

[18] Eussen SJ, Ueland PM, Clarke R, Blom HJ, Hoefnagels WH, van Staveren WA, et al. The association of betaine, homocysteine and related metabolites with cognitive function in Dutch elderly people. The British Journal of Nutrition. 2007;**98**(5):960-968. DOI: 10.1017/S0007114507750912

[19] Li H, Lu H, Tang W, Zuo J. Targeting methionine cycle as a potential therapeutic strategy for immune disorders. Expert Opinion on Therapeutic Targets. 2017;**21**(9):861-877. DOI: 10.1080/14728222.2017.1370454

[20] Berardis DD, Orsolini L, Iasevoli F, Tomasetti C, Mazza M, Valchera A, et al. S-Adenosyl-L-methionine for major depressive disorder. In: Melatonin, Neuroprotective Agents and Antidepressant, Therapy. New Delhi, India: Springer; 2016. pp. 847-854

[21] Román GC, Mancera-Páez O, Bernal C. Epigenetic factors in late-onset alzheimer's disease: MTHFR and CTH gene polymorphisms, metabolic transsulfuration and methylation pathways, and B vitamins. International Journal of Molecular Sciences. 2019;**20**(2):319. DOI: 10.3390/ijms20020319

[22] Tehlivets O, Malanovic N, Visram M, Pavkov-Keller T, Keller W. S-adenosyl-L-homocysteine hydrolase and methylation disorders: Yeast as a model system. Biochimica et Biophysica Acta. 2013;**1832**(1):204-215. DOI: 10.1016/j.bbadis.2012.09.007

[23] Lee ME, Wang H. Homocysteine and hypomethylation. A novel link to vascular disease. Trends in Cardiovascular Medicine. 1999;**9**(1-2):49-54. DOI: 10.1016/s1050-1738(99)00002-x

[24] Paganelli F, Mottola G, Fromonot J, Marlinge M, Deharo P, Guieu R, et al. Hyperhomocysteinemia and cardiovascular disease: Is the adenosinergic system the missing link? International Journal of Molecular Sciences. 2021;**22**(4):1690. DOI: 10.3390/ijms22041690

[25] Durand P, Prost M, Loreau N, Lussier-Cacan S, Blache D. Impaired homocysteine metabolism and atherothrombotic disease. Laboratory Investigation. 2001;**81**(5):645-672. DOI: 10.1038/labinvest.3780275

[26] Škovierová H, Vidomanová E, Mahmood S, Sopková J, Drgová A, Červeňová T, et al. The molecular and cellular effect of homocysteine metabolism imbalance on human health. International Journal of Molecular Sciences. 2016;**17**(10):1733. DOI: 10.3390/ijms17101733

[27] Jong CJ, Sandal P, Schaffer SW. The role of taurine in mitochondria health: More than just an antioxidant. Molecules. 2021;**26**(16):4913. DOI: 10.3390/molecules26164913

[28] Franco R, Schoneveld OJ, Pappa A, Panayiotidis MI. The central role of glutathione in the pathophysiology of human diseases. Archives of Physiology and Biochemistry. 2007;**113**(4-5):234-258

[29] Clemente Plaza N, Reig García-Galbis M, Martínez-Espinosa RM. Effects of the usage of l-cysteine (l-Cys) on human health. Molecules. 2018;**23**(3):575. DOI: 10.3390/molecules23030575

[30] Maclean KN, Jiang H, Phinney WN, Mclagan BM, Roede JR, Stabler SP. Derangement of hepatic polyamine, folate, and methionine cycle metabolism in cystathionine beta-synthase-deficient homocystinuria in the presence and absence of treatment: Possible implications for pathogenesis. Molecular Genetics and Metabolism. 2021;**132**(2):128-138. DOI: 10.1016/j.ymgme.2021.01.003

[31] McCaddon A, Regland B. COVID-19: A methyl-group assault? Medical Hypotheses. 2021;**149**:110543. DOI: 10.1016/j.mehy.2021.110543

[32] Dattilo M, Fontanarosa C, Spinelli M, Bini V, Amoresano A. Modulation of human hydrogen sulfide metabolism by micronutrients, preliminary data. Nutrition and Metabolic Insights. 2022;**15**:11786388211065372. DOI: 10.1177/11786388211065372

[33] Rehman T, Shabbir MA, Inam-Ur-Raheem M, Manzoor MF, Ahmad N, Liu ZW, et al. Cysteine and homocysteine as biomarker of various diseases. Food Science & Nutrition. 2020;**8**(9):4696-4707. DOI: 10.1002/fsn3.1818

[34] Kitada M, Ogura Y, Monno I, Xu J, Koya D. Effect of methionine restriction on aging: Its relationship to oxidative stress. Biomedicine. 2021;**9**(2):130. DOI: 10.3390/biomedicines9020130

[35] Safrhansova L, Hlozkova K, Starkova J. Targeting amino acid metabolism in cancer. International Review of Cell and Molecular Biology. 2022;**373**:37-79. DOI: 10.1016/bs.ircmb.2022.08.001

[36] Potęga A. Glutathione-mediated conjugation of anticancer drugs: An overview of reaction mechanisms and biological significance for drug detoxification and bioactivation. Molecules. 2022;**27**(16):5252. DOI: 10.3390/molecules27165252

[37] Safrhansova L, Hlozkova K, Starkova J. Targeting Amino Acid Metabolism in Cancer. New York, NY, United States: Elsevier BV; 2022

[38] KS MC. Review: Chemical pathology of homocysteine VI. Aging, cellular senescence, and mitochondrial dysfunction. Annals of Clinical and Laboratory Science. 2018;**48**(5):677-687

[39] KS MC. Chemical pathology of homocysteine VII. Cholesterol, thioretinaco ozonide, mitochondrial dysfunction, and prevention of mortality. Annals of Clinical and Laboratory Science. 2019;**49**(4):425-438

[40] Moshal KS, Tipparaju SM, Vacek TP, Kumar M, Singh M, Frank IE, et al. Mitochondrial matrix metalloproteinase activation decreases myocyte contractility in hyperhomocysteinemia. American Journal of Physiology. Heart and Circulatory Physiology. 2008;**295**(2):H890-H897. DOI: 10.1152/ajpheart.00099.2008. Epub 2008 Jun 20

[41] Belal C, Ameli NJ, El Kommos A, Bezalel S, Al'Khafaji AM, Mughal MR, et al. The homocysteine-inducible endoplasmic reticulum (ER) stress protein Herp counteracts mutant α-synuclein-induced ER stress via the homeostatic regulation of ER-resident calcium release channel proteins. Human Molecular Genetics. 2012;**21**(5):963-977. DOI: 10.1093/hmg/ddr502. Epub 2011 Nov 1

[42] Pushpakumar S, Kundu S, Sen U. Endothelial dysfunction: The link between homocysteine and hydrogen sulfide. Current Medicinal Chemistry. 2014;**21**(32):3662-3672. DOI: 10.2174/0929867321666140706142335

[43] Quinzii CM, Lopez LC. Abnormalities of hydrogen sulfide and glutathione pathways in mitochondrial dysfunction. Journal of Advanced Research. 2020;**27**:79-84. DOI: 10.1016/j.jare.2020.04.002

[44] Harbison RA, Ryan KR, Wilkins HM, Schroeder EK, Loucks FA, Bouchard RJ, et al. Calpain plays a central role in 1-methyl-4-phenylpyridinium (MPP+)-induced neurotoxicity in cerebellar granule neurons. Neurotoxicity Research. 2011;**19**(3):374-388. DOI: 10.1007/s12640-010-9172-4

[45] Hannibal L, Blom HJ. Homocysteine and disease: Causal associations or epiphenomenons? Molecular Aspects of Medicine. 2017;**53**:36-42. DOI: 10.1016/j.mam.2016.11.003

[46] Ostrakhovitch EA, Tabibzadeh S. Homocysteine and age-associated disorders. Ageing Research Reviews. 2019;**49**:144-164. DOI: 10.1016/j.arr.2018.10.010

[47] Robinson K. Homocysteine, B vitamins, and risk of cardiovascular disease. Heart. 2000;**83**(2):127-130. DOI: 10.1136/heart.83.2.127

[48] Shiao SPK, Lie A, Yu CH. Meta-analysis of homocysteine-related factors on the risk of colorectal cancer. Oncotarget. 2018;**9**(39):25681-25697. DOI: 10.18632/oncotarget.25355

[49] Nandi SS, Mishra PK. H_2S and homocysteine control a novel feedback regulation of cystathionine beta synthase and cystathionine gamma lyase in cardiomyocytes. Scientific Reports. 2017;**7**(1):3639. DOI: 10.1038/s41598-017-03776-9

[50] Ganguly P, Alam SF. Role of homocysteine in the development of cardiovascular disease. Nutrition Journal. 2015;**14**:6. DOI: 10.1186/1475-2891-14-6

[51] Carey A, Fossati S. Hypertension and hyperhomocysteinemia as modifiable risk factors for Alzheimer's disease and dementia: New evidence, potential therapeutic strategies, and biomarkers. Alzheimer's and Dementia. 2022;**19**(2):671-695

[52] De Chiara B, Sedda V, Parolini M, Campolo J, De Maria R, Caruso R, et al. Plasma total cysteine and cardiovascular risk burden: Action and interaction. The Scientific World Journal. 2012;**2012**:303654. DOI: 10.1100/2012/303654

[53] van den Brandhof WE, Haks K, Schouten EG, Verhoef P. The relation between plasma cysteine, plasma homocysteine and coronary atherosclerosis. Atherosclerosis. 2001;**157**(2):403-409. DOI: 10.1016/s0021-9150(00)00724-3

[54] El-Khairy L, Vollset SE, Refsum H, Ueland PM. Plasma total cysteine, pregnancy complications, and adverse pregnancy outcomes: The Hordaland homocysteine study. The American Journal of Clinical Nutrition. 2003;**77**(2):467-472. DOI: 10.1093/ajcn/77.2.467

[55] Aghayan SS, Farajzadeh A, Bagheri-Hosseinabadi Z, Fadaei H, Yarmohammadi M, Jafarisani M. Elevated homocysteine, as a biomarker of cardiac injury, in panic disorder patients due to oxidative stress. Brain and Behavior: A Cognitive Neuroscience

Perspective. 2020;**10**(12):e01851. DOI: 10.1002/brb3.1851

[56] Perna AF, Ingrosso D. Atherosclerosis determinants in renal disease: How much is homocysteine involved? Nephrology, Dialysis, Transplantation. 2016;**31**(6):860-863. DOI: 10.1093/ndt/gfv409

[57] Perna AF, Ingrosso D, Satta E, Lombardi C, Acanfora F, De Santo NG. Homocysteine metabolism in renal failure. Current Opinion in Clinical Nutrition and Metabolic Care. 2004;**7**(1):53-57. DOI: 10.1097/00075197-200401000-00010

[58] Kruglova MP, Ivanov AV, Fedoseev AN, Virus ED, Stupin VA, Parfenov VA, et al. The diagnostic and prognostic roles played by homocysteine and other aminothiols in patients with chronic kidney disease. Journal of Clinical Medicine. 2023;**12**(17):5653. DOI: 10.3390/jcm12175653

[59] Kruglova MP, Ivanov AV, Virus ED, Bulgakova PO, Samokhin AS, Fedoseev AN, et al. Urine S-Adenosylmethionine are related to degree of renal insufficiency in patients with chronic kidney disease. Laboratoriums Medizin. 2021;**52**(1):47-56. DOI: 10.1093/labmed/lmaa034

[60] Shih YL, Shih CC, Chen JY. Elevated homocysteine level as an indicator for chronic kidney disease in community-dwelling middle-aged and elderly populations in Taiwan: A community-based cross-sectional study. Frontiers in Medicine (Lausanne). 2022;**9**:964101. DOI: 10.3389/fmed.2022.964101

[61] Dandge VA, Variya D. Study of vitamin B12 deficiency in chronic kidney disease. International Journal of Advances in Medicine. 2020;7:303-307

[62] Capelli I et al. Folic acid and vitamin B12 administration in CKD, why not? Nutrients. 2019;**11**(2):383

[63] Kowluru RA. Diabetic retinopathy: Mitochondria caught in a muddle of homocysteine. Journal of Clinical Medicine. 2020;**9**(9):3019. DOI: 10.3390/jcm9093019

[64] Wijekoon EP, Brosnan ME, Brosnan JT. Homocysteine metabolism in diabetes. Biochemical Society Transactions. 2007;**35**(Pt 5):1175-1179. DOI: 10.1042/BST0351175

[65] Wang JX, You DY, Wang HP, et al. Association between homocysteine and type 2 diabetes mellitus: A systematic review and meta-analysis. International Journal of Diabetes in Developing Countries. 2021;**41**:553-562. DOI: 10.1007/s13410-021-00933-9

[66] Li C, Qin J, Liu W, Lv B, Yi N, Xue J, et al. Profiling of homocysteine metabolic pathway related metabolites in plasma of diabetic mellitus based on LC-QTOF-MS. Molecules. 2023;**28**(2):656. DOI: 10.3390/molecules28020656

[67] Holeček M. Role of impaired glycolysis in perturbations of amino acid metabolism in diabetes mellitus. International Journal of Molecular Sciences. 2023;**24**(2):1724. DOI: 10.3390/ijms24021724

[68] Gupta S, Kühnisch J, Mustafa A, Lhotak S, Schlachterman A, Slifker MJ, et al. Mouse models of cystathionine beta-synthase deficiency reveal significant threshold effects of hyperhomocysteinemia. The FASEB Journal. 2009;**23**(3):883-893. DOI: 10.1096/fj.08-120584

[69] Park EJ, Je J, Dusabimana T, Yun SP, Kim HJ, Kim H, et al. The uremic toxin

homocysteine exacerbates the brain inflammation induced by renal ischemia-reperfusion in mice. Biomedicine. 2022;**10**(12):3048. DOI: 10.3390/biomedicines10123048

[70] Lehotský J, Tothová B, Kovalská M, Dobrota D, Beňová A, Kalenská D, et al. Role of homocysteine in the ischemic stroke and development of ischemic tolerance. Frontiers in Neuroscience. 2016;**10**:538. DOI: 10.3389/fnins.2016.00538